AF615168

Growing Food in South Florida

Tropical Gardening Series

1 Growing Food in South Florida

Felice Dickson

Growing Food in South Florida

BANYAN BOOKS
Miami, Florida

Designed by Bernard Lipsky

Manufactured in the United States of America
by Miami Book Manufacturing, Inc., Miami, Florida

Library of Congress Cataloging in Publication Data

Dickson, Felice, 1943-
Growing food in south Florida

[Tropical gardening series: 1]
Bibliography: p.
Includes index.
No. 1. Vegetable gardening—Florida. 1. Title.
SB21.5.D52 635.09759 75-31739
ISBN 0-916224-00-7

NOTE: Due to continuing review by the federal Environmental Protection Agency, all pesticide recommendations in this book should be considered current as of publication date only. Check with your County Cooperative Extension Service for changes.

Second Printing, 1976

Contents

Illustrations & Charts

Illustrations

Charts

Thanks are due to the following for sharing their advice and expertise on home vegetable growing: Dr. William Stall, Dade County Cooperative Extension Service Specialist—Vegetables, who read the book in manuscript; Lewis Watson, Broward County Cooperative Extension Service agent; Dr. John Popenoe, director of the Fairchild Tropical Garden; the late Alexis Bervy, a careful backyard "chemical farmer;" organic growers John Dickler and Ira Ebersole; and Laurace Dickson, as always.

Growing Food in South Florida

An Introduction

When people came to South Florida in the early part of this century, they came to stay. As the furniture was uncrated, memories of how things used to be done were packed away in the knowledge that this unique region had another way of life in store.

Discovering this new way of life was no hardship in a land that tossed delicious foods back at gardeners as fast as they could plant the seeds. As put by the late Mabel Dorn, an enthusiastic horticulturist and horticultural historian of the area, there was never a lack of familiar and exotic food year-round in this "botanical Eldorado" where the people lived on the land and off of its products.

Then came the boom years that brought thousands of newcomers who sent word to all of their relatives and friends that Florida had become entirely civilized, with stores full of apples and black suits. The wave of new settlers was looking for warm weather, not adventures in farming, which most had happily left behind. And the "close-in" land became more valuable to build on than to grow on.

Many left after the Depression, broke and with no further use for land that had become worthless. After World War II, the boom resumed, the price of land rose, and today we find even acreage with an Everglades view valued so dearly that land speculators are almost the only ones who can afford to farm it.

The planting Eldorado lives on, however, in every South Florida backyard, and with the rising power rates and high food

Everyone in the family enjoys home vegetable growing. This is a raised plot, also called a "frame garden," bordered by coral rock and filled with topsoil and well-composted stable manure. String marks straight rows for a second planting of turnips.

prices in northern states, residents in warm weather areas are beginning to realize its golden glow has not tarnished with urbanization. Every piece of property is a potential year-round mini-farm, regardless of how much concrete surrounds it, regardless of its rocky, sandy surface.

"Nearly everyone in the early pioneer days was plant-conscious and developed the soil and its products with genuine joy and profit," said Dorn, and it looks as though those "good old days" are here again. Plant your own eggplants and chayotes. Eat your own broccoli and bananas. Practice makes perfect peppers, and in the bargain you'll find you're eating better, exercising more, gaining a finer appreciation of South Florida—an incredible and unequaled oasis of livability in these prepackaged times.

The Tropical Difference

Many are the farmers—both backyard and pro—who were successful in northern climes but who moved to warm-weather areas and were reduced to rank amateurs. When they attempted to grow vegetables again, nothing happened.

A long growing season sounds like heaven to anyone who wants to raise his or her own food, but learning to feed yourself in the tropics requires a resolution not to attempt to do it the "easiest" way or the old way but the best way for this latitude.

Prolonged warm weather—and especially mild weather punctuated by cold snaps as in South Florida—changes the entire growing picture from that found in a strictly temperate zone. It means more bugs, a greater potential for fungus and other diseases, and a new calendar for the division of vegetables into fall, winter, spring, and summer growers.

If you are a "chemical" gardener, it means more spraying and soil fumigation. If you are an organic gardener, it means greater diligence and more patience. If you are somewhere in between, it calls for living a little closer to your patch, making it more a part of your life than just a means to an end. And that's the fun of it.

This book is written for all three types of gardeners. Its goal isn't to show how hard mild-weather vegetable growing can be. The growing is different but that doesn't mean difficult. Getting off to the right start will save you all of the retrenching and patchwork common to beginners, and you'll wonder—as you lug in your first bushel basket of homegrown produce—what so satisfying could be as simple.

To give the tropics equal time, a section is included on growing the kinds of vegetables that "should" be grown in warm-weather areas, as opposed to the "temperate-zoners" that American palates favor. (Hard-to-find cooking directions are included for those who would like to try these "natural" foods.)

Accent the Negative?

Everyone knows how things go when they go right. You make a wish, and the wish comes true. You want to plant a cabbage, you plant it, you feed it, you water it, it grows, you eat it.

Sometimes, however, things don't run that smoothly. You wonder what went wrong, how you could have fixed it, what else might go wrong, and how you could prevent it.

This book attempts to cover the significant problems gardeners would be likely to encounter in a very unlucky season. The overwhelming majority of these setbacks, however, would be only temporary or would never crop up for the vegetable grower who prepared the plot correctly and stuck to a regular, simple spraying schedule.

The emphasis on spraying may distress organic gardeners; in fact, most people are looking for a way to grow more naturally. So, rather than devote a separate chapter to organic growing, natural methods–when they exist–will be offered as alternatives when potential problems are discussed.

A few problems a grower may run across are not his or her fault but can be blamed on the weather or other "invisible" factors. These, too, will be covered.

Get an Agent

Singers, writers, spinach growers–they all do better with an agent. In the vegetable growers' case, the one they need is their county agent, an employee of the Florida Cooperative Extension Service.

The county agent is a university-trained expert in ornamental trees and plants, vegetables, fruit, and other subjects. In a small county he may be just one man; in a larger county, there's a man for each specialty.

The Cooperative Extension Service is administered by a university–in Florida, the University of Florida–and funded by federal, state, and local governments. The university prints up free booklets on hundreds of subjects and has the county agent distribute them through his office, by phone or mail request. He also distributes U.S. Department of Agriculture publications. In addition, the agent himself answers all questions put to him by the public.

"The public" isn't just a professional farmer with a lot of money at stake in his crop, but may be one gardener with four cucumbers. County agents are paid to be bothered by the most amateur questions as well as the most complex. If you think you have a "stupid question" on vegetables, or a complex one, call it in to your friendly county agent. He will think you're smart for taking it to an expert instead of to an understaffed garden shop or to a neighbor.

County agents are found by looking for the listing "Cooperative Extension Service–Agricultural Division" under county offices or below the subhead "Department of Agriculture" under U.S. government offices.

First Things First

The importance of planning to the success of your first—or your fiftieth—vegetable plot cannot be overemphasized. If you are not the methodical type, resolve just this once, at least, to give careful planning a try because there is nothing so fatal to the enthusiasm for growing things as 100 square feet or more of blighted, bug-ridden produce.

If your background in vegetable growing is scant, you will have to assume a vigilance unfamiliar to the average gardener. The short life of vegetables as compared to grass, trees, and shrubs means that irreversible damage can occur much more quickly than with most landscaping materials. If you only examine your cantaloupe as infrequently as you do your citrus, you may be too late to apply first aid.

The following steps are the first moves for consideration before you're actually ready to get your hands dirty:

1. Order your seed catalogs.
2. Select your site.
3. Select your seed.

Why Seed Catalogs?

No one contemplating a vegetable garden of any variety can count on finding the best types locally, either as seedling plants or as seed. While a garden shop may have the right variety of eggplant, peppers, and maybe one of the best tomatoes, it's unlikely that it will have much more in "liners," those light plastic six-packs of young plants. In addition, many people come to grief by not knowing how to select seedling vegetables. They look for the huskiest, biggest plants instead of the smallest

When buying vegetable seedlings, check on root development. The smaller the plant, the better, providing color is good. Plant at left shows maximum root development; plant at right is ideal.

plants with good color—which is what they should be taking home.

Seedling plants in liners may look robust, but the root systems of these youngsters are very soon damaged by both the small size of their containers and the lack of nutrients. Water and light from above may keep their foliage looking good, but if the roots are crowding the sides (lift a prospective purchase out a little and take a look), the maturing plant will never reach the size of a seedling transplanted at the right time. This does not mean a seedling has to be root-bound before it starts to suffer permanent root damage but the roots will be very visible on all sides.

The selection of seed in garden shops is handicapped by two factors: the wrong varieties for the tropics and subtropics are often arbitrarily sold to garden shop owners, few of whom know the difference, and the viability (life) of the seed is reduced by too high humidity. Air conditioning is not adequate to maintain viability. Most garden shop personnel are familiar only with the maintenance of grass and perennials and have no idea of the vast differences that exist between growing vegetables in the temperate zone and in the borderline tropics.

This is not to say that a certain number of people who buy their seed at garden shops where the trade is brisk and the stock is fresh do not have good-looking plots. It is simply a cautionary piece of advice to gardeners who do not always feel lucky. If you decide to buy in haste strictly from what's available and have trouble later, consider this as the beginning of your problems.

The three seed companies most frequently patronized by South Floridians are the George W. Park Seed Co., Inc., Greenwood, South Carolina 29646, the W. Atlee Burpee Co., Philadelphia, Pennsylvania 19132, and the Kilgore Seed Company, 1400 West First Street, Sanford, Florida 32771. Also among the leading dealers are Burgess Seed and Plant Co., P.O. Box 218, Galesburg, Michigan 49053 and R. H. Shumway, Seedsman, 628 Cedar Street, Rockford, Illinois 61101, the latter especially for unusual varieties.

Park has the flashiest catalog and is well known for the pains it takes with its seed and reputation. It is famous for its "make-goods." Burpee is old, reliable. Kilgore offers many of the South Florida-recommended seeds. Shumway is America's oldest seedsman.

The reason you should order more than just one catalog is that no single seed house may have all of the South Florida-recommended varieties. Parks, for example, carries 'Homestead No. 24' and 'Manalucie' but not 'Tropic,' a fine staking tomato. Burpee carries 'Tropic' but not the other two.

Be sure you request the latest vegetable catalog as opposed to the current one which, in late summer, may carry mostly bulbs. Waiting for your catalogs to arrive is not the waste of time it appears to be to eager-beavers. There is plenty to do meanwhile, both on paper and in the preparation and fumigation of the plot site. The latter requires a two-week wait before planting, anyway.

Selecting the Site

The number one mistake of beginning gardeners is locating the vegetable plot where it is most convenient instead of where it gets full sun. Full sun is necessary for all vegetables that bear fruit. Leafy vegetables can squeeze by with partial shade. The definition of the terms "full sun" and "partial shade" are frequently misinterpreted.

"Full sun" means that at any time of the day the sun has a clear path to the spot in question. You can cheat on this all-day definition by bringing it down to six hours, but if less than that your yield will be noticeably reduced. Four hours of direct (full) sun would constitute enough for leafy vegetables. Or they could have six to eight hours of sun-dappled light such as that provided by a high oak.

The path of the sun during the prime winter growing season should be considered, rather than the amount of sun remembered in a spot during the summer. A tree or fence that was no problem during one season could cast a much longer shadow at another time of the year.

If there is any question about how much sun a prospective location receives, have someone spot-check the light every couple of hours or prune any trees in question. Almost invariably, the vegetable gardener who decides to go ahead with a questionable shade situation regrets it.

How Big?

Unless you have an unlimited amount of sun, the amount you do have will limit how much you can grow. So space available would be the first consideration in determining how large your plot can be.

There is a school of thought that believes a first-timer should not attempt to provide all of a family's vegetables and, instead, should select two or three favorites and "practice" on these. If you can contain your enthusiasm or want to test your determination to make growing at least part of your food a way of life, this might be a provident way to start. A 20-by-40-foot patch takes very little more time to care for than a 10-by-10, but the investment in soil (if needed), fertilizers, and sprays is considerably less, of course, for the smaller plot.

And, if you hope to can your surplus but aren't certain about how much work that involves, a mini-plot will let you practice without a lot of waste. (With a freezer, time-consuming canning is no longer a question; for free booklets on canning and freezing, see Selected References, pp. 123-4.)

Hesitant beginners may like the idea, too, of starting out with a mini-plot and supplementing their harvest with trips to the "u-pick-em" fields. These are the vegetable fields already picked for harvest but which have enough "stragglers" in them to make profitable opening them to public picking. (See directions to Dade County's "u-pick-em" fields in Appendix.)

If, however, you know you are going to make a go of it sooner or later, you might as well "go the whole route," and before soil and fertilizers get any higher in price, gain your experience with a plot large enough to provide most of your vegetable intake.

Now back to how big. For several years, Dade County agents pointed to the backyard Miami garden of a retired New York state dairyman, Alexis Bervy, as a model small-scale, grow-all-you-need vegetable plot. Bervy kept such careful records that his plot was picked by Time-Life Books to feature as 10 pages of color in *Vegetables and Fruits,* a volume of its Encyclopedia of Gardening. (See Selected References, p. 124.)

Bervy said his 22-by-42-foot plot was planned to feed a family of four (himself, his wife, a daughter, and her child), but it actually could provide food year-round for 10 people or more. Here's what he planted in that space, in rows running east and west, starting at the south end.

Bervy's Choices

CUCUMBERS: One row of 40 'Poinsett' replanted twice in six-months' time. They mature in three months and are picked two or three times a week until the end of the harvesting season in six months.

Yield per row each planting was 120 pounds with 12 quarts of pickles preserved.

CABBAGES: Three rows of 'Early Jersey Wakefield,' 22 plants each row. Seeds planted continuously replaced harvested heads. Heads mature in three months and are picked weekly over a six-month season.

Yield per row per planting was 22 heads, with eight quarts of sauerkraut preserved.

PARSLEY: One row with 22 plants of 'Dwarf Green Curled.' Plants matured in about three months, but some sprigs could be picked in two. One planting produces enough for garnishes over a six month period.

CARROTS: 'Chantenay' is the variety he preferred, planting one row and replanting as needed. Seedlings were thinned to 250 plants an inch apart. They matured in about three months. The Bervys used two plantings within six months.

The yield was 45 one-pound bunches per planting.

LETTUCE: He made three successive plantings of two rows of 22 plants of the variety 'Great Lakes,' 10 inches apart. Early heads were ready in about three months, and mature heads were picked over four months.

The yield was 22 heads per row per planting.

BEETS: 'Detroit Dark Red' was seeded in two plantings of about 120 plants per row, two inches apart. They were ready in two months and were picked weekly during a four-month harvesting season.

The yield per row per planting was about 27 bunches, weighing about a pound and a half each. Ten quarts were preserved.

ONIONS: Two rows of Southport Yellow Globe' and one row of 'Sweet Spanish' were planted. One planting was made of each. They were grown from seed and thinned to about 100 plants per row. Both varieties were ready in about three months and were harvested over a six- to seven-month period.

The yield was 30 pounds per row.

EGGPLANTS: One row of six plants of 'Black Beauty,' were set 18 inches apart. They started to yield after two and a half months. Harvesting continued for more than six months.

The yield was 15 to 25 eggplants, each one to two pounds, with 12 quarts of eggplant relish preserved.

BROCCOLI: The variety was 'Calabrese'—one row of 10 plants, 18 inches apart. They matured in about three months. Heads were picked two or three times a week over seven months.

SNAP BEANS: 'Harvester' was chosen. Seeds were sown in groups every two weeks, totaling one row of 120 plants spaced two inches apart. The beans matured in two months. As they stopped bearing, they were replaced by new seed. Harvest season is seven months.

Twenty-five pounds was the yield, with 20 one-quart packages frozen.

GREEN ONIONS: Green or bunching onions are immature onions planted as "sets" or small bulbs. If they are not picked, they grow on to mature for use as ordinary large onions. Bervy planted a row of about 200 onion sets and picked them starting in six weeks.

The yield was 20 pounds or 64 five-ounce bunches.

SWEET PEPPERS: One row of 12 'California Wonders,' spaced about 18 inches apart. They began maturing in three months, produced for six months.

Bervy's yield was about 450 peppers with 10 pints pickled.

TOMATOES: 'Manalucie' was his preferred variety. He planted a row of 15 and replanted twice over a six-month period. The first tomatoes matured in three months. Each planting produced for two months. Ten pounds were picked twice a week.

Yield was about 75 pounds, with 10 quarts of whole tomatoes preserved and 10 pints of chili sauce made.

SQUASH: Two clumps (or hills) of three or four hybrid zucchini plants were sown, one on each side at opposite ends of the plot. They began to mature in about two months and were picked over a seven-month period.

The yield was about 75 squashes per hill.

More Recommended Varieties

Don't let a tempting seed packet in a garden shop or a catalog make you believe that just because you're looking at it, it will grow in South Florida or other tropical areas. Several familiar northern vegetables such as globe artichokes, rhubarb, and asparagus grow poorly here or require special handling not practical for most gardeners.

In addition, many varieties of vegetables that will grow well here are not suitable for subtropical-tropical farming. There are other varieties, instead, that have been hybridized for their resistance to common warm-weather problems. Failure to investigate what types grow best in warm climates has resulted in legions of drooping tomato vines and other "mysteriously" diseased vegetables.

The lack of prolonged cold weather makes insects and dis-

Which vegetables grow best in South Florida? Selection of the most disease- and weather-resistant varieties can be as important as choosing the vegetables that do best in warm climates. With a list of the recommended hybrids, ordering from seed catalogs is less confusing.

eases more of a problem here since they can survive in the soil all year, but University of Florida agricultural researchers have found varieties that will stand up to them. (See Planting Guide for Vegetable Gardens in Appendix.) They are available from the seed companies previously listed.

As you are planning your plot, note which vegetables should be planted only at certain times of the year, and check the approximate length of harvest for each plant. This will make possible a closer guess as to the number of successive plantings you will need to keep your favorite vegetables on the table.

A Season for All Things

A very common stumbling block with first-timers is rushing into the ground a vegetable that should wait until either warmer or colder weather. English ("garden") peas should not be planted before the first of November in South Florida and not later than the first of February for the production of well-filled pods.

Okra is a warm-climate and long-daylight-requiring crop that should not go into the ground between October and March. The remaining months of the year will grow this vegetable satisfactorily, providing that the soil is not too heavily infested with nematodes. Okra is an easy mark for these pests—probably one of the most susceptible.

Cautious growers and nongamblers plant their first vegetables around the first of October in South Florida. By this time—and by the time the tiny plants start to appear—the nights will have begun to cool and the seedling-flooding rains common to summer largely are past. A few gardeners start in September but they usually are the ones who have enough free time to run out and sprinkle their panting little plants at least twice a day.

In February or March the last push is made to plant cool-weather seeds so that they will mature before the nights start warming up in May. This February-March planting gives gardeners a second chance to remedy the mistakes they made during their first season or to try out different varieties, as well as to extend the growing season.

Starting in March, growers also put in the vegetables they hope to harvest during the summer. These warm-weather types include pole beans, squash, okra, Southern (black-eyed) peas, collards, cucumbers, and peppers. Onion sets will work, too.

Gardeners interested in vegetables native to the tropics like chayotes and malangas can plant these anytime. Most are perennials. You'll find more information on them beginning on p. 83.

What Will It Cost?

In the past, making a profit by growing your own vegetables was unlikely in South Florida. Cost of spraying, fertilizing, soil-building, watering, and canning made a gardener lucky to break even.

With the inflation following the first dramatic rises in vegetable prices in 1973, however, home grocery growing began to show a profit. In 1974, the National Garden Bureau compiled the following cost comparison lists for seed for a 15-by-25-foot garden as compared to the same vegetables purchased at what many might consider a cut-rate market for the times. The Bureau's garden featured 18 varieties of vegetables yielding a continuous supply of fresh produce for nine months, with plenty left over for a family of four to can or freeze.

Total cost of the Bureau's seed was $9.30. For that sum, it bought $284 worth of vegetables.

Additional costs in South Florida might be four yards of topsoil (about $48) if you are on rock and want a raised plot, a 100-pound bag of 6-6-6 (30 per cent organic) for about $8, an eight-ounce jar of combination insecticide-fungicide for about $1.60, and a quart bottle of Vapam ($2.89) or a four-ounce bottle of Nemagon ($1.35) for soil fumigation, plus any tools or sprinkling devices you might want. Any way you add it up, you come out way ahead. (If topsoil is needed, it is an investment that can be spread over many seasons.) Some of these prices may already be obsolete but they are indicators.

Here is the Bureau's cost comparison of market food to seed:

Yield from 15x25 Foot Plot	Market Value	Cost of Seed
60 Cucumbers @ 25c each	$ 15.00	$ .50
100 lbs. Tomatoes @ $1.00 for 3 lbs.	33.00	.50
40 lbs. Zucchini @ 39c per lb.	15.60	.50
40 lbs. Peppers @ 39c per lb.	15.60	.50
24 heads Cabbage @ 39c head	9.36	.35
48 heads Lettuce @ 49c head	23.52	.35
25 lbs. Beans @ 39c lb.	9.75	.35
48 lbs. Chard @ 59c lb.	28:32	.35
36 lbs. Beets @ 29c lb.	10.44	.35
36 lbs. Carrots @ 29c lb.	10.44	.35
12 lbs. Spinach @ 59c lb.	7.08	.35
24 bunches Radish @ 29c bunch	6.96	.35
48 bunches Parsley @ 29c bunch	13.92	.35
24 bunches Green Onions @ 25c bunch	6.00	1.20
28 bunches Leeks @ 59c bunch	16.52	.35
24 heads Broccoli @ 49c head	11.76	.75
12 heads Cauliflower @ 79c head	9.48	.75
15 lbs. Peas @ 39c lb.	5.85	.75
60 pts. Brussels Sprouts @ 59c pt.	35.40	.35
TOTAL	$284.00	$9.30

Preparing Your Plot

While you're waiting for your seed to arrive, you can shape up your plot and make sure both you and the young plants will start off the season with every possible advantage. If you have plenty of sun, choose the spot closest to a water spigot and farthest from trees and shrubs whose roots will compete with your vegetables for nutrients and water. If the garden must be located near some definite threats, dig a ditch up to a foot or more deep along the side where roots could encroach, and place roofing paper or a sheet of heavy plastic film (available at larger hardware stores) along one side of the trench. Refill the trench with soil to form a barrier to roots.

Grass that may be covering the proposed site of the vegetable patch does not necessarily need to be removed. Some people break up the sod and turn it over. Others cover it with newspapers, put soil over the papers to hold them down, and smother it for a month before they are ready to begin the bed. Then they put their topsoil over the papers and just let the papers decompose underneath as mulch.

Selecting the Soil

In South Florida, there are well over a dozen kinds of soils but for gardeners, essentially three different types: sand, marl and rock. Of these, marl is to be preferred, not nutrition-wise (they are all quite deficient) but because it holds moisture—and therefore fertilizers—longer than the others. This is the soil that covers many of the newer subdivisions bordering the Everglades. It is removed when the acreage is drained, filled, and graded, and it is sold in nurseries as "black dirt" or topsoil.

Although virtually all soils in South Florida that are not marl appear to be sandy, some sand soils are much deeper than others. Deep sand (over a foot in depth) is convenient for vegetable-raising and can be made more moisture-retentive by the addition of peat moss or compost (see p. 49 for composting instructions). This organic matter should be added so that water and fertilizer will be retained by the soil. It is usually mixed up to a half-and-half ratio as deep as six inches. One-third peat moss to two-thirds soil is also a great improvement. The expense

Vegetables can be grown in native Dade County rock, as the mammoth farming belt here proves annually, but it is an expensive process based entirely on chemical culture. Most Dade Countians prefer starting from scratch with topsoil or adding organic matter to existing land.

of peat might be considered a necessary evil the first season, but by the second time around, the compost pile should be producing enough free humus.

Deceptively sandy on the surface is the soil that covers a large section of south Dade County. It may be only three to six inches deep in some areas. Underneath is oolite—oolitic limestone, a rock unusual in the United States but common in the Bahamas, the Florida Keys, Western Cuba, the Yucatán Peninsula, and elsewhere. Geologically, it is a very interesting material, but to the gardener it looks hopeless. Still, scarified as it is by the farmers of Dade, it produces gigantic harvests. Or rather, it "holds up" some fine-looking vegetables (serving much the same use as gravel in hydroponic gardening) while fertilizers, water, insecticides, and fungicides—plus the mild climate—are responsible for the harvest.

Rototilling is not popular among home gardeners on these rocky soils, since it is much simpler to construct a very simple

Two kinds of borders were chosen for the popular raised beds shown here. Pressure-treated wood can be used as a simple frame or can be made more elaborate as an attractive garden feature. Cement blocks are quickest to install. Creosoted telephone poles are also used.

raised or "frame" bed for small-scale farming. The plot is outlined with eight-inch concrete blocks, telephone poles, railroad ties, coral rocks, stacked creosoted posts, or treated wood boards (the "frames" of the term frame garden) and filled with topsoil. The soil depth gained in this way makes it possible to grow better-sized root crops like carrots and beets. Also, better root development is encouraged for other crops, watering needs to be done less often, and fertilizing is necessary less frequently.

Alternatives to Topsoil

Some people may find topsoil initially expensive. There are other ways to fill your "frame," especially if you can haul a large quantity of material. Local riding stables often will strike a deal with gardeners so that they can cart away loads of manure or manure mixed with wood shavings. The manure must be thoroughly composted (rotted) before it is a suitable planting medium. If you live near a racetrack, call the grounds supervisor. He may let you take away all you need at no charge.

Another source of free bedding mix is the local sewage treatment center. Treated dry sewage sludge does not have an odor problem. It does have a tomato seed problem since these seeds survive both human and chemical processing. The only way you'll eliminate stray seeds in sludge is to treat your plot first with a fumigant like Vapam or a pre-emergence herbicide (both requiring a waiting period before planting) or learn to recognize tomato seedlings and pull them as they appear.

The Fumigation Question

The straight-and-narrow path cannot be beaten for coming as close to total success as is humanly possible. And at the top of the list of straight-and-narrow procedures for vegetable growers in South Florida is fumigation of plots to kill nematodes, soil-borne diseases like damping-off on young seedlings, and weed seeds.

This does not mean that for several seasons you can not raise vegetables on land that has not been fumigated, despite what fumigation enthusiasts may say. After the first couple of seasons, your crops may be stunted and not produce well, but if you keep the soil fertility level high and plant more, you still will have something to show for your efforts—except, perhaps, with a crop like okra, which will not tolerate nematodes.

Nematodes are microscopic, eel-like organisms that do various kinds of puncturing and sucking damage. Some of their handiwork shows up as knots on the roots, but nematodes other than the root-knot nematode do other kinds of damage, resulting in dieback, fruit or bud drop (also caused by other factors), loss of leaves, or overall loss of vigor. The most common sympton is failure of the plants to develop beyond a certain point.

You can buy sterilized soil that will make this no concern of yours—for awhile. And most topsoil is collected at a depth that is almost nematode-free. So, with both sterilized soil and new topsoil, the first few crops should cause no worry about these pests. They do, however, find their way into new soils eventually, so if you want to keep yours relatively "clean," it would be better to use a nematicide or soil fumigant before the soil becomes well infested. The second year would be a wise time to start, but the first season would be even wiser.

Commercial Growers Have the Edge

This is an appropriate time to give away the "secret" of commercial growers who produce thousands of acres of near-perfect-looking plants without coddling each one, as a mini-farmer sometimes feels he or she must. (Even so, commercial growers do inspect their fields daily.)

In addition to being able to use spray materials more potent than homeowners are allowed to buy (because, in theory, they are too dangerous for the amateur to use), commercial growers are allowed to use a much stronger soil fumigant, one easily fatal in untrained hands.

These restricted chemicals smooth the way considerably for the pro, but this doesn't mean amateurs have to sulk because they can't eliminate problems as efficiently. Not every leaf in nature is perfect, and if your crop is not counted down to the last leaf, you should have more than enough for your needs.

How to Use Fumigants

Two fumigants are by far the most widely used by backyard farmers. They are Nemagon and Vapam. Nemagon and similar products kill nematodes only. Vapam is more expensive and kills weed seeds, nematodes, and damping-off organisms.

Nemagon can be applied as a granular or liquid material in furrows four to six inches deep and 12 inches between furrows. In liquid form, it goes on in a 50 percent form at the rate of three tablespoons in just enough water to drench 100 feet of row or, if the 30 percent granules are used, four ounces for 100 feet of row.

After Nemagon is applied, the furrow should be covered and the plot soaked and kept wet for 48 hours. After 48 hours the soil should be uncovered and allowed to dry out so that the Nemagon gas can escape. Two weeks should be allowed for this aeration. Nemagon is not recommended as a fumigant where peppers, eggplants, and onions will be grown, so grow these "on the side."

A quart of Vapam will treat 100 square feet. It is diluted with water and applied to the soil with a sprinkling can or hose-end sprayer. Soak thoroughly, then cover with a plastic sheet. The sheet has to be held down around all edges with cement blocks, pipes, or what-have-you to keep the fumes inside. It is left in place for two days.

There is a two-week waiting period, also, with Vapam before the site is ready to plant. Both products have complete instructions on their containers.

Alternatives to Fumigation

Applying a fungicide like Captan or Maneb to each row of seed or the entire bed can be effective in varying degrees in the prevention of damping-off, a problem that shows as the sudden collapse of young seedlings. There is no waiting period.

Damping-off is hard to control since it is caused by a soilborne fungus. You should try to keep some residue of fungicide in the soil beginning with the plants' emergence when you should spray or dust. Then, three applications made once or twice a week should get seedlings past the critical point.

Many people have attempted to affect nematodes by planting

marigolds in their vicinity. It has been proven that marigolds' roots exude a compound toxic to nematodes, but a dainty border of these plants is not enough. They could be used as a cover crop in between seasons or plantings, or they could be alternated in rows between the rows of vegetables as well as around the border of the plot.

Mulching between rows may also be effective in holding down nematode populations. The materials used should be quick to break down because it is this process of breaking down that discourages the pests, not the mere presence of mulch. A mulch of pine needles or bark would not be satisfactory.

Another way to control the nematode population is to maintain a high fertilizer level without, of course, damaging the plants. The salts in the fertilizer act as a dessicant, dissolving them. Light but frequent fertilizing with a 6-6-6—perhaps half a coffee can per 30 feet of row every other week—should do it, but experiment first on a test section of your plot. Fertilizer burn shows as a marginal browning or overall yellowing of the leaves, and it is easy to lose vegetables quickly with a too-heavy application of fertilizer. (A fertilizer with 100 percent organic nitrogen and as many other components as possible derived from organic sources is easier on the plants.)

Complete Control Isn't Necessary

Unless the infestation in your yard is very heavy (which is quite likely in older gardens), it is possible for you to achieve an adequate harvest of vegetables not especially sensitive to nematodes if there are enough nutrients provided to encourage extra strong growth.

Nematodes may stunt the growth of vegetables not grown in treated soil, but with extra attention to watering and fertilizing, you can produce enough in spite of them. Overplanting would be a good idea, too, in this case. (That doesn't mean crowding, just planting extra amounts.)

Nematode-Sensitive Plants

The following vegetables are especially sensitive to nematodes: okra, chayotes, rozelle, squash, potatoes, bell peppers, Southern peas, eggplants, carrots, beets.

Refrigerating Seeds and Sets

After your seeds arrive, you may want to make provision at that time for storage or, in some cases, pre-planting chilling. Pepper seeds, in particular, should be stored in a screw cap jar in the lower part of a refrigerator for at least two weeks before they are planted. This chilling treatment results in faster germination and sturdier plants.

Lettuce seed, too, benefit from this chilling. You'll find that seed stored in a refrigerator for a year will germinate faster than it did as fresh seed. But, of course, this doesn't mean you have to wait a year before you plant your first order.

Onion sets should be stored in a refrigerator for two or three weeks before planting. They should not be in a closed container but in a well-ventilated tray or bag since they must "breathe" or they will dry out. Leftover seed can be saved in screw cap jars on a refrigerator's lower shelves, and it will remain usable for one to three years.

Treated Seed

Many vegetable seeds are treated with a chemical to prevent damping-off and attack by soil-borne disease. Seed potatoes are frequently treated with a chemical to break the dormancy of newly-dug potatoes. Cut seed potatoes should be treated with Maneb soon after they are cut to reduce seed piece rot after they are planted.

None of these treated seeds should be eaten, and any that are stored in the refrigerator should be clearly marked "Poison–Do Not Eat" and kept away from children.

Growing from Seed in Flats

There is no one best way to sprout seeds and to grow plants successfully. Seeds can be sown directly into a plot and later thinned as recommended, or they can be raised in flats, then thinned and transplanted. One South Florida gardener likes to grow his in topsoil tossed into an old wheelbarrow.

Usually the determining factor is the time of year. Some gardeners who want to get a head start on the season that begins in October grow in flats, thereby enabling them to protect seedlings from the torrential rains and hot sun common to late summer and early fall.

Those who use Nemagon instead of Vapam and want to protect young plants against damping-off without the further expense of post-emergence fungicides often elect to sprout in flats, using a sterile medium such as two parts peat moss to one part perlite.

The handiest containers are those trays made of plastic that are used by nurserymen to hold plastic six-packs of seedlings. These can be purchased at better garden shops.

Any other kind of plastic tray (with holes punched in the bottom for drainage) or wooden flats will do. Some people use egg cartons or ice cube trays. To decrease the chance of their harboring soil-borne organisms, they should not have contained soil before, although many growers take a chance, anyway, washing them out thoroughly beforehand (this does not sterilize them, of course).

The best place to put these flats is under a tree giving medi-

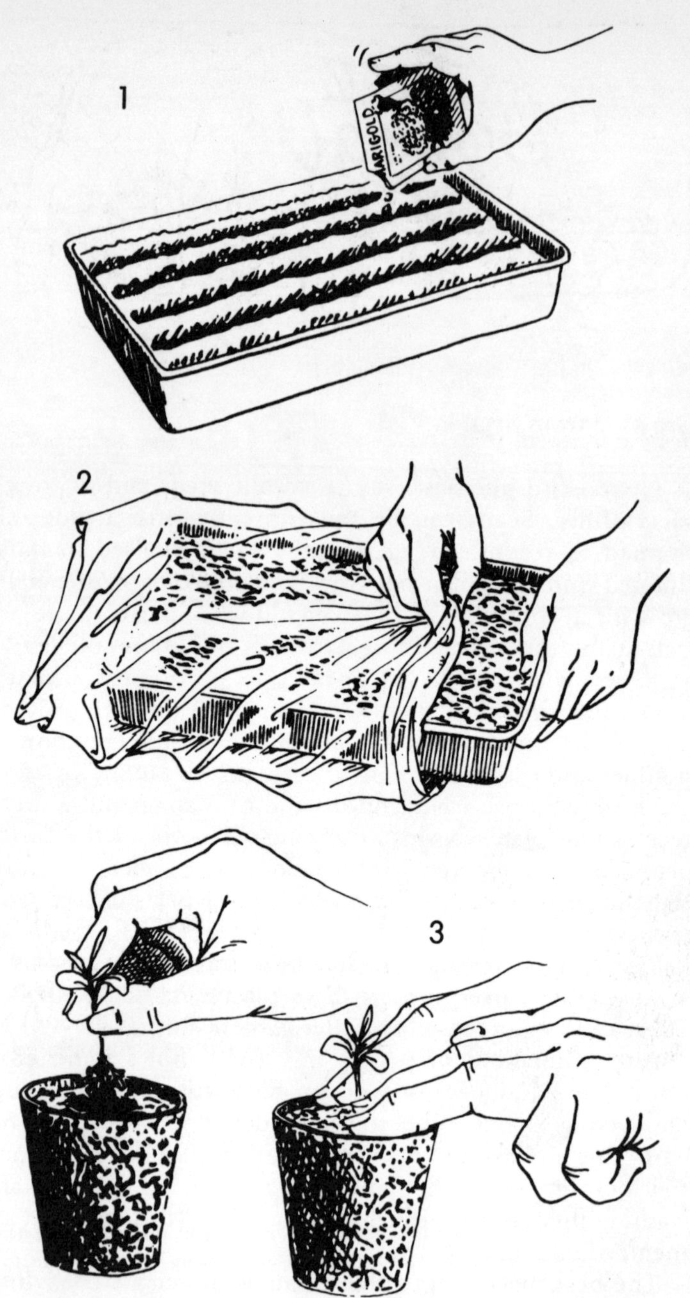
1
MARIGOLD
2
3

Planting seeds out in the garden too early in the season can result in washouts from heavy rains. Here are two methods of raising transplants from seed in protected areas: (1) Sow seed in dampened sterilized soil or potting mix in plastic tray, water is lightly, and cover it (2) with a plastic bag. Set in light shade. When second set of leaves appear, transplant (3) to three- or four-inch peat or plastic pots. (4) Cover with plastic bag. In two weeks, without extra watering, most should be ready for setting out.

Seed also can be sown in special planting blocks, watered well, and stored in a tray covered with plastic. When leaves appear, open the bag but keep blocks moist to the touch. About two weeks later, set entire block out in garden.

um to light shade. They can also be set in the sun if they are misted regularly. A favorite device for shading the flats somewhat and for helping repel hard-hitting raindrops is a jalousie door or a smaller window screen, propped over the flat so there is an inch or so of ventilation on at least one side.

A piece of glass or Plexiglass could be used too. These would contain the moisture better than the screening but ventilation should be allowed.

Seeds are sown according to instructions on the package or in the planting guide. Most beginners plant seeds deeper than necessary. Tiny seeds can be scattered across the soil surface and settled in with a good spraying. Small seeds like carrots are planted shallowly and fairly closely together. They help each other break through the soil. Some prefer to plant these little ones by cutting off a corner of the seed pack and tapping it gently to urge them out.

Seed catalogs offer seeders ranging in size from little more than an eye dropper to row seeders that can be walked. Some seeds come covered with a seed protector. If you are concerned about damping-off because you haven't taken other precautions, you can dust on a seed protector like Arasan (thiram) yourself.

Moisture for Seedlings

A few vegetables require more moisture than others when tiny. Celery likes the most, followed by beets and lettuce.

Seed flats should not be kept soggy (which encourages damping-off problems) but should not be allowed to dry out. Peat moss, a fairly reliable indicator of the moisture level, turns a light brown as it dries.

Temperature for Seedlings

As indicated before, temperature is critical to successful sprouting with some vegetables. Preferring cool weather are English peas, cabbage, lettuce, and onions. Seeds that sprout better in warmer weather are cucumbers, muskmelons, okra, bell peppers, and watermelons.

A winter garden should not be planted prior to the middle of

September, and continuous planting can be made, depending upon space available, until the first of April.

If transplants are used for tomatoes, peppers, and eggplants, the seeds should have been planted not earlier than the middle of September. This would mean that the plants would not be available before mid- to late October. If tomatoes, eggplants, and pepper seeds are planted when the soil temperature is above 90 degrees F. at night, as well as in the daytime, the first several clusters or individual flowers will not produce fruit.

Temperatures affect the pollen of their flowers six to eight weeks before the flower buds appear. High temperatures and temperatures below 40 degrees F. will damage the pollen so that desirable fruit will not be produced.

Damping-Off: The First Mystery

Gardeners are so elated to see seeds sprout and grow that when their tiny seedlings suddenly topple over, the inexperienced are shocked. Usually the stems of the plants near the ground appear to have been pinched. This is a classic symptom of damping-off, caused by one of several fungi. There is also pre-emergence damping-off, in which sprouting seeds are attacked underground.

Damping-off is encouraged by rainy, cold weather that retards the growth of plants and makes them more susceptible to the disease. Seeds planted too deep and a planting mix containing green mulch also encourage it.

Treated seeds help hold it back, plus a spray program beginning on emergence. Spray them with Captan, Maneb, or Fermate once or twice a week. If you plan to use a combination insecticide-fungicide (discussed later) for regular control throughout the season, you may elect not to purchase a fungicide separately but to start right out with the combination material.

Damping-off is hard to control once it has entered your seedbed, but you can lessen its impact by drenching with a fungicide.

Food for Seedlings

After your seeds come up they are hungry and should be fed

weekly with a water-soluble 20-20-20 fertilizer mixed at half strength with water. (A fertilizer with a higher first number [nitrogen] is desirable.)

Transplanting Seedlings

The first pair of leaves to appear are not true leaves, but by the time the second pair appear, you can consider transplanting. You can use a pencil to urge them gently from the soil, holding the stem cautiously with your other hand. Transplants can go into six-packs, other trays, or peat Jiffy pots set in trays filled with peat-perlite or a combination of peat moss, perlite, and sterilized soil. (It is probably easier and cheaper to buy a small bag of sterilized soil at a garden shop than to treat it or sterilize it for a couple of hours in your oven.)

When the plants are three or four inches high, they are ready to go into the plot. Water the seedling and the surrounding soil thoroughly. If you take the transplant from a shady situation to full sun overnight there is danger of burning the young plant, so move shaded transplants into partial sun for a few days first. (Hardest to transplant or not worth the effort because of a short season to harvest are squash and cucumber relatives, beans, corn, turnips, radishes, and carrots.)

Seeding in the Row

Seeds frequently are planted so deep that the emerging plant expends all of its energy forcing the growing shoot above the soil surface. This mistake is made with small seeds as well as large. All seeds, regardless of size, should be planted according to instructions on their package or by using the following guidelines:

Large seeds such as beans, peas, and squash may be planted about an inch deep. Smaller seeds like carrots, radishes, and onions should be planted no deeper than a quarter-inch. Seeds of squash, peppers, tomatoes, and similar crops that are to be spaced at considerable intervals in the row may be planted directly in the soil. After the seeds have emerged, the plants can be thinned to a desirable spacing, leaving the most vigorous plants to grow on.

Tomato seeds may be planted in a row. When the plants are four to six inches high, the extra plants may be pulled and transplanted to another row, thereby extending the harvest season of the crop. Peppers and eggplants can be handled this way, too.

Radish seeds should be sowed thinly in the row, and extra plants should be removed so that the distance between plants is about an inch and a half. Rows should be at least 18 inches apart. Radishes grow rapidly and can be interplanted with carrot seeds to mark the rows. As carrots are slow to emerge, the radish crop will be harvested before the carrots need the space. Carrot seedlings should be thinned to about two inches between plants. If radishes or carrots are planted too thickly in the row, their tops will develop with very little, if any, root development.

Laying Out Your Plot

In designing your plot, several points should be considered. Taller growing plants should be on the north side of the garden since the rows are going to run east and west to take maximum advantage of the winter sun.

Rows should not be so closely spaced and plants in the rows should not be so crowded as to cut off sun, as well as intensify competition for nutrients. Despite all the ready information on seed packs and planting charts, many gardeners, for one reason or another, cheat on space allotments. Whether it is an attempt to grow too many in too small a space or a reluctance to thin young plants, it does not pay.

Too-closely-spaced root crops will develop foliage with little or no root development. Tomatoes will be stunted, also. Plants that are being grown for the tops only can be spaced more closely—following recommendations on the seed pack.

Most backyard plots are small enough to trellis tomatoes, cucumbers, and pole beans for maximum production. These trellises can be made by using sticks or fencing supported by posts at either end of the rows. A six-inch mesh reinforcing material is fine. Trellises should be about five to six feet high since the varieties of tomatoes, pole beans, and vine crops normally grown in South Florida will reach at least this height.

If if is within your budget and your yard isn't fenced, it's a good idea to surround the plot with a do-it-yourself fence high enough to keep out dogs, poultry, rabbits (if they are a factor where you live), and children. The fence not only protects the garden but serves as a trellis for pole beans, tomatoes, and other crops needing support. It can be as simple as chicken wire fastened to wood posts set with concrete into halved sections of concrete blocks.

A fence makes it easy to achieve straight rows by measuring off the correct row distances on one side with paper clips, twist-ems, or whatever, corresponding distances on the opposite fence, and connecting the points with string. Then you follow alongside the string with a stick, yardstick, or other narrow row-making device.

You can also make straight rows by jamming sticks in the ground at the appropriate distances and connecting them with string. People using treated board frames sometimes mark off their top edge with nails.

Crops should be grouped by similar planting and maturing dates if you want to keep the entire garden in production as much as possible. Then, within each grouping, consider arrangements according to the size of the plant.

Arrange low-growing vegetables such as radishes, turnips, mustard, and lettuce along one side of the plot, the medium-tall plants such as peppers, bush beans and squash in the middle of the garden, and the tall vegetables along the other side to minimize shading.

More than one crop, each requiring about the same spacing, may be planted within a single row. Under certain conditions, planting on the south side of a bed running east and west might have advantages in wind protection and quicker soil warm-up over sensitive vegetables.

Crops that span more than one season, such as strawberries, should be placed to one side of the garden so that they do not interfere with seasonal preparation of the garden.

Interplanting quick-growing crops like radishes among slower-growing ones is a common practice. The fast-growing radishes are out of the way before the longer-growing crop needs the space.

Allow ample space between rows for convenient cultivation and harvesting, even if some plants like turnips or Bibb lettuce may need rows marked off only a foot apart. If you're clumsy—and have enough room—you may appreciate a little more space as they mature.

Don't plant too much of any one crop at a time, especially those crops that must be eaten fresh, like radishes, and cannot be stored. To provide fresh vegetables over a long period of time, make interval plantings of any one vegetable every 10 to 14 days. This practice works particularly well for crops such as beans, sweet corn, and peas, which have a short peak period of quality.

Plant two or more varieties having different maturity dates to prolong the season for any one crop. While genetic crossing may occur, this is a problem only in sweet corn where "xenia" effects show up on the ears (example: yellow kernels mixed with white ones).

A popular mistake with corn is to plant one or even two rows. Sweet corn should be planted in blocks rather than in single rows so that plenty of pollen is present in the air around the corn stalks. The blocks, of course, do not need to be a full row long. Figure on how much corn you would like to have and divide it into three, four, or more shorter rows, if short rows are all you need. This practice should produce better pollination and ear fill-out.

Weed Control

Some gardeners prefer to take measures to avoid weeds even before they plant seed, or seedlings. One of these precautions is fumigation with Vapam. Another is laying down rolls of black plastic, a common practice in the commercial growing of strawberries in South Florida. The plastic is cut to the length of the rows and held down with rocks, stakes, or any other device suitable to the task.

Half the fertilizer the plot will need for the season can be laid down before the plastic is put on. The rest is applied through the planting hole and, if desired, through random holes made.

Plastic commonly has been used for lettuce production also. It is obtainable at many garden shops and hardware stores. Only people with a strong distaste for weeds extend its use throughout the vegetable plot.

Most growers who live where lawns are kept mowed find

weeds a minor problem. They usually are infrequent enough to remove by hand if they are not allowed to go to seed. Cultivation with hoes and rakes—almost a trademark of home vegetable gardening elsewhere—should not only be unnecessary but should be avoided since a too-vigorous application can damage small feeder roots that are near the surface.

Nutgrass is a serious weed problem with no real control. It will puncture plastic and is not affected by most of the herbicides that are available.

Herbicides, by the way, are not recommended for vegetable gardens since many of the materials that are available do not have Environmental Protection Agency clearance for use on food crops, a federal requirement to protect people as much as the environment.

Mulching is a very simple way to reduce or eliminate weeds. If you do see a weed in mulch, it is usually very spindly and easy to pull. Pine needles make a good mulch. So do small leaves. Grass clippings may be full of weed seed unless your lawn is well-cared-for. Put on a good thick layer between rows.

During the summer, if the vegetable plot is not in use, you can keep weeds away and add organic material to the soil by applying grass clippings and leaves to the area. Some gardeners like to grow a cover crop of sorghum or an inexpensive legume. Spade the cover crop into the soil about four weeks before planting in the fall.

Cutworm and Mole Cricket Control

Before your plot has a chance to show what it can do, it may fall victim to damage by these two pests. Laying down one of the many baits made for these bugs before seeds or seedlings are planted can thwart the mysterious cutting-in-two of young vegetables. Mole crickets can cause devastating root damage that a young plant can't afford.

Both of these insects rarely show themselves, so gardeners are often baffled by the cause of their damage. Bait should be scattered across the plot. Edging the plot with it is another precaution, especially if the bait cannot be used near a particular crop.

Fertilizing

There are almost as many recipes and schedules for fertilizing as there are growers—and more than one approach is the "right" way. But one thing is certain about South Florida soils; they need almost everything added except lime.

On the positive side, this makes understanding what you will have to add to your soil quite simple. There is no real need to worry about pH (which hovers around 7.0 for most of South Florida).

Several approaches will be given here. All are being used by gardeners who are entirely satisfied with their harvests. Choose the one that appeals to you, and if you think you could be getting more from your plants, switch to another next season.

Method A. Grower A uses one 100-pound bag of 6-6-6 (30 percent organic) on a 20-by-30 plot, and it meets all the nutrient needs for a season. He puts half of it down when he plants, the rest in two- to three-week intervals after the seeds sprout. His soil is sand with peat moss added.

Method B. Grower B digs in Milorganite according to instructions on the bag when he prepares the bed. After the plants have two sets of leaves he starts fertilizing lightly every two weeks with a side-dressing (between-row fertilizing) of 6-6-6. His soil is marl topsoil.

Method C. Grower C uses composted manure mixed with straw from a local stable. He fertilizes "occasionally" with a 4-8-8.

Method D. Grower D uses a 6-6-6 (100 percent organic nitrogen). He sets out his seedlings and at that time puts down a coffee can of fertilizer per every 30 feet of a row as a side dressing. Three weeks later he puts down half a can every 30 feet, repeating the amount three weeks after that. He gives radishes and small plants less than larger, longer-growing vegetables, and gives lettuce not as much as pole beans.

In the free Extension Service booklet, "Vegetable Gardening Guide," prepared by retired Dade County Agent Nolan Durre, another approach is offered. If you are using 6-6-6, take a pound coffee can full and distribute it evenly for 100 lineal feet of row in a furrow four inches deeper than the placement of the seeds or plants. This is to give the fertilizer enough time to go into the soil solution before the roots enter the fertilizer zone.

For side dressing crops that will be growing for more than 60 days, half the amount of fertilizer recommended for planting should be applied on either side of the row, six inches from the plants, four weeks after seeding.

For a long-growing crop, a second application in the same amount should be applied six weeks after the first side dressing. Durre suggests that the fertilizer not be mixed into the soil since normal moisture and irrigation of the garden will dissolve the fertilizer and carry it into the root zone to feed the plant as the plant requires it.

Some gardeners like to take advantage of the odor-free, dry treated sludge available free from their friendly sewage treatment plant. This has the possible disadvantage of containing tomato seeds that are not affected by human or mechanical processing, so if you see a seedling out of place, you should pull it since—unless it was grown by a knowledgeable gardener—it is probably the wrong variety.

Sewage sludge, stable manures, and Milorganite-enriched soils are very high in nitrogen, very low in almost everything else. Nitrogen (the first number on a fertilizer bag) is valuable to the growth of young plants, but as they approach vegetable adolescence they benefit from equal amounts of phosphorus (the second number) and potash (the third).

A phosphorus deficiency can delay maturity and cause plants and fruit to be smaller. Fruit may not appear at all. Potash makes stems stronger, plants more resistant to disease, and fruiting better.

Some people using natural fertilizers topheavy with nitrogen bring their crops to harvest with no complaints, but most prefer to add a chemical fertilizer somewhere along the line. Their choice should be something similar to a 4-8-8 since a 6-6-6 will contribute more nitrogen than the plants need. Plants do not take what they need from the soil and leave the surplus; the elements will enter the root systems essentially in the same proportions as exist in the soil. The so-called secondary elements such as iron, magnesium, manganese, and zinc are also necessary, but they are present in most chemical fertilizers.

Organic fertilizers have many plus features. They do not cause fertilizer burn and serve as buffers when inorganics are added. They are natural chelating agents—making more easily available secondary elements that are hard for some plants to assimilate in high-pH soils.

They make soil more "friable," looser and more receptive to rain. They hold other fertilizers better. They are longer-lasting.

On the negative side, they have to be applied, generally, in large quantities to be effective. In 100 pounds of cow manure there are only a half pound of nitrogen, one-third pound of phosphorus, and a half pound of potash. In 100 pounds of 6-6-6 fertilizer, there are six pounds each of nitrogen, phosphorus, and potash. It takes 1,200 pounds of cow manure to equal the same amount of major elements found in 100 pounds of 6-6-6.

There are many sources of inexpensive or free organic nitrogen; organic phosphorus and potash are a little more difficult to obtain. The table on p. 120, prepared by University of Florida researchers, shows the relative value of various sources of organics.

Manures vary considerably in their content, the composition varying according to type, age, and condition of animal; the kind of feed used, the age and degree of rotting of the manure, the moisture content, and the kind and amount of litter or bedding mixed in the manure.

Here is a guide to the rate of application:

Cow, horse, and hog manure: 25 pounds per 100 square feet (about five tons per acre) of garden soil. For best results, supplement each 25 pounds of manure with two to three pounds of superphosphate. (Organic gardeners may want to substitute raw bone meal.)

Poultry and sheep manure: 12 pounds per 100 square feet (about three tons per acre), supplemented with one to two pounds of superphosphate or raw bone meal.

After planting, manures can be applied as sidedressings—up to five pounds per 100 square feet of row for cow, horse, and hog manure; three pounds per 100 square feet of row for poultry and sheep manure.

If you are also using a mulch, rake it back at the edge of the root zone to apply the band of manure, then re-cover with the mulch.

Are Organics Better than Inorganics?

Some of the merits of organics were just discussed, and if gardeners favor them additionally for reasons of economy, ecology, or aesthetics, they are on firm ground.

If, however, they feel that their vegetables will be nutritionally or visually superior through the use of organics, or that chemical or inorganic fertilizers have nutritionally negative characteristics, they will not find confirming evidence among recognized agronomists. They say that the soil converts all organic material to chemical components indistinguishable and, indeed, identical in composition to chemical fertilizers, and it is only in this chemical form that they can be assimilated by roots, anyway.

Secondary Element Deficiencies

Vegetables can show a deficiency in one or more of the secondary (other than nitrogen, phosphorus, and potash—the "big three") elements. Better fertilizers will contain these elements in adequate amounts, but cheaper fertilizers may not.

Magnesium sulfate, either in the fertilizer or applied in small quantities to all crops, will be of benefit, in addition to manganese. Manganese sulfate helps beans, strawberries, and certain other crops. A deficiency of magnesium and manganese is indicated by fading color. When manganese is deficient, the veins of the leaves will remain green. A five percent manganese dust can be used for a quick recovery from manganese deficiency.

Corn can show a zinc deficiency with pale leaves, light green stripes, and often a deformed appearance. Zinc sulfate can be applied as a spray or in the fertilizer. Or it can go into the soil separately at planting time, when it should be applied six ounces per 1,000 square feet.

If you are working a small plot, it is easier to take a list of these essential elements with you when you go shopping for a fertilizer and read the "Sources Derived From" sections of the label—as well as the secondary listing itself—to get one that makes this backtracking unnecessary.

The Care and Feeding of Compost Piles

A manure-like organic fertilizer—"artificial manure"—can be created by composting all of your grass clippings, leaves, and table scraps not containing meat.

Compost is made by alternating layers of these organic materials with manure, topsoil, lime, and chemical fertilizer (optional), and combining them with water and air.

Most compost piles are not less than 10 feet square and three to five feet high. Those most often seen in South Florida are bordered on three sides by cement blocks or fencing (sometimes with a gate on the fourth side) so that the pile is open to be turned frequently.

A compost pile definitely is not a trash pile to be abandoned. The latter is the kind of heap that gives compost piles a bad name as a fly or pest attractor. Compost piles should be turned at least every three weeks to keep air inside the pile. This is necessary to keep aerobic bacteria in residence. It is this bacteria that is responsible for the breaking down of the organic matter.

Between turnings, the top of the compost pile should be left flat or with a slight depression in the center to catch rain or added water. (Too much water eliminates air and slows the decay process.)

One way of the several in which a pile can be built is to make a layer of leaves, straw, grass clippings, or other organic materials a foot deep, wet it down and pack it. Then spread a layer of manure four to six inches deep over this layer of wet material. Next spread up to three pounds of superphosphate and a pound of ground limestone.

Instead of the superphosphate, you could use five pounds of an inexpensive 6-6-6. The limestone can be omitted or a shovelful of South Florida limestone soil substituted.

Repeat the process until the pile reaches the desired height.

The compost will begin to heat after two or three days. Keep it moist, but not too wet, until it's time to fork it over, mixing the parts well. You can add to this rich mix pine straw, fish scraps, water hyacinths, and the neighbors' castoff bags of grass clippings.

A compost pile turns leaves and clippings into what is missing in South Florida soil—rich humus. One side of this bin is hinged to swing open for frequent turning of its contents; this is necessary to speed up decomposition of the organic matter.

The best location for the pile is a sunny or partly sunny spot. Compost should be ready in about two months if the weather has been warm and the pile kept mixed. When it is broken down into a homogenous mixture and no undecomposed leaves or other material are distinguishable, it is ready for use.

Use it as you would manure, broadcasting it over the entire garden at the rate of about 25 pounds per 100 square feet or a quarter pound per square foot. If you want to put down only a small quantity at a time it may be mixed into the soil alongside each planting furrow or at each hill site.

Watering

The standard rule of green thumb for watering vegetable gardens is to put down at least an inch every week. If the skies provide all or part of this requirement, fine. But you'd better have a rain gauge—even if it's only a marked glass—on a cement block in the center of your plot to make sure you're not guessing wrong.

A soaking with one-half to one inch of water twice weekly during the warmer parts of the season and once weekly during cool weather is a good recommendation to follow. Thorough soaking of the ground is better than a light sprinkling every day or so. An indication that your plants aren't getting enough water is if they remain wilted at sunset.

Young seedlings of plants such as lettuce may need a midday sprinkling if the weather is unseasonably warm. If you can't be around at this hour, listen to weather reports and water well in the morning on a day expected to be warm.

Various opinions exist as to the importance of keeping water off the plants. Obviously this is impossible when it rains, and commercial growers water with overhead sprinklers. If you spray regularly, you should give no thought to it (although watering in the late afternoon is not recommended). If you are interested in trying to make it organically, you might want to try flooding in your water with a bubbler on the end of your hose and the rows set out on ridges so that the water can run down beside each row. This will prevent organic insecticides from washing off the plants unnecessarily.

Spraying for Disease and Pest Control

The emphasis in problem control on tropical and subtropical crops is frequent, regular spraying (not dusting, because you want to coat the undersides of leaves, too).

There are different approaches and combinations of approaches that the backyard gardener can take, but by far the easiest is using a combination insecticide-fungicide every five to seven days—every three to five days during wet, cool weather.

These sprays, sold as water-soluble powders, may be called

"Tomato and Vegetable Spray" or just "Vegetable Spray," but a close inspection of the label will show further labeling as a "fungicide with insecticides" and a listing of the ingredients—most often Maneb and zineb for disease control, Diazinon for aphids, small caterpillar, plant-bugs, and leaf miner control, and Sevin for worms. (Sevin is fatal to bees so use with care.)

The label will also show how close to harvest the material should be used for safety; usually they should not be used within three days of harvest on tomatoes or within seven days on other crops. These sprays cannot be mixed ahead of time and should be stored in a cool, dry place.

A combination insecticide-fungicide such as the one described will control late blight, early blight, anthracnose, downy and powdery mildews, frogeye spot and alternaria spots, horn worms, pin worms, fruit worms, most grasshoppers, mole crickets, potato beatles, bean beetles, pumpkin bugs, thrips, serpentine leaf miners, army worms, aphids, and stink bugs.

Occasionally you might feel the need for an insecticide that kills something not covered by your combination spray. You may want, for instance, to use Kelthane, a miticide that destroys both mites and their eggs. This probably is the only other product you might want if you have a good combination spray.

If you buy and use fungicides separately, you should remember that some have exclusive uses and some can be harmful. Karathane is used only to prevent powdery mildew, while Benomyl or Benlate prevents powdery mildew plus most other fungus diseases but not downy mildew. Neutral copper is an old fungicide that is persistent and should not be used frequently on soft leaved plants such as cucumbers and melons.

Regular use is especially important with fungicides since they are effective only as preventatives on areas of plants that have not been invaded by a fungus spore. Application should start as soon as plants are out of the ground and continue as long as the plant is growing.

Coverage is important with sprays; a hit-and-miss application has caused many people to complain that a spray didn't work. The coarse spray from a bottle gun or hose-end proportioner isn't adequate (except in controlling lawn problems). A Flit gun is awkward and won't let you get to the undersides of leaves.

(Plus the fact that it won't spray wettable–water-soluble–powders well.)

Trombone-type sprayers take so much pumping they can wear you out, but there are some small plastic bottle sprayers with well-made brass hand pumps that budgeters with perseverance can use to advantage.

The best sprayer is a compression (tank) type. You mix your spray material in a bucket, pour it in (don't mix it in the tank), pump up the compression, then pull the trigger and spray until it's time to pump up the compression again. (The spray material should be sloshed around in the tank occasionally to keep it in suspension.) Compression sprayers are available in a wide price range, and if you are committed to growing your own vegetables, you will never regret buying the best that is most resistant to the corrosive action of harsh chemicals.

Organic Controls

The emphasis placed in South Florida on spraying vegetables disturbs many people who feel that a more natural way should be possible. Complicating the picture, however, is an unnatural factor–the desire to grow vegetables not native to an area whose natural balance between predators and controls has been widely upset.

Add to this the difficulty many working people have in being able to keep close tabs on their plots and to hand-pick bugs, and it is apparent they will be tempted frequently to resort to chemical control. Most spray materials recommended in this book are on a list distributed by Tropical Audubon Society as among those less persistant in the environment. For those who want to stick closer to nature, however, there are alternatives.

If bugs are not too numerous, worms can be controlled by hand-picking. (Picking diseased leaves will not control them since the spores that develop in and on a diseased plant are present on the plant before the effects are noticed.)

Baccillus thuringiensis is a well-known organic remedy, safe to use on vegetables for chewing insects. It is a microbial insecticide harmless to humans and animals and does not require a waiting period before harvest. It is sold under the names of Dipel, Biotrol, and Thuricide. It kills caterpillars (such as cabbage loopers) only.

Rotenone is an old standard, safe to use right up to harvest time. It controls bean leaf beetles, maggots, cabbageworms, and many other chewing insects. It is often combined with the natural daisy powder, pyrethrin (which kills aphids but is also fatal to ladybugs, so be careful).

Perma-Guard is a nonpoisonous garden insecticide, harmless to children, food, and pets. It gives immediate and lasting control of most insects that are killed by physical action, not by poisoning.

An even simpler physical action is to use the strongest stream of water from your hose to knock off mites and break their webs. Aphids and leafhoppers can be shooed off in this way, also, although this method would be ineffective with a massive infestation. Beetles or larvae can be knocked off by hand into a jar or pail of water coated with a film of oil or kerosene.

Some organic gardeners start seeds or seedlings under wax paper plant caps pushed an inch into the ground to keep out egg-laying beetles. For cutworms, they set seedlings inside a waxed milk carton, tar paper, or just paper collars (often inverted paper cups, bottoms slit, pushed an inch into the soil) to prevent the cutworms from reaching the plants.

Organic gardeners in the tropics need to be much neater than most. Nearby debris and ornamental plants infested with a problem should be removed and destroyed, as should all infected vegetable plant parts. After a harvest, plant debris also should be removed.

Organic growers do well to buy sterilized soil and to start their seeds indoors in pasteurized potting soil or in sterile peat pellets. When setting out plants, they should avoid crowding, overwatering, and overfertilizing seedlings.

Some organic gardeners will use lime and "simple" sulfur (both "natural" chemicals). Sulfur will deter mites, aphids, and cutworms (but don't overdose or use in hot weather). It can be mixed with a little lime to dust young cabbages and broccoli. A mild mix discourages aphids and worms.

An early dusting, followed by regular applications, of sulfur before mildew gets underway is a preventative, but keep in mind that sulfur can burn foliage or fail to act at all, depending upon its fineness and temperature. An old remedy for leaf miners is a spray made from hot pepper juice.

Various light traps can be made to attract the night-flying moths of cutworms and armyworms. A Coleman lantern mounted above a large pan of waste oil is one. An ordinary hundred-watt bulb placed between a fan and a removable bag is another contraption often called a "Florida bug trap." Both mealybugs and scale insects can be controlled by soapy water, but this definitely is not appropriate for an infestation.

Making a garden attractive to lizards, birds, and frogs will benefit your vegetables but there is some doubt as to the value of buying praying mantis, ladybugs, and lacewings. They have a tendency to move on, sometimes before they finish the job at hand and certainly if a continuous diet is not present.

There are many new books—many in paperback—on organic gardening, and while little of the cultural information applies in South Florida, some of the pest and disease control information may be very useful.

A Bug Overview

Healthy vegetables not allowed to wilt or become undernourished are very resistant to the depredations of bugs, and routine spraying will make them a small concern. Even so, it's useful to know what you're up against.

Aphids. All vegetable gardeners experience aphids, also known as plant lice. They come in many species, usually green and very tiny. The leading aphid pests are bean, cabbage, melon, potato, pea, and turnip aphids.

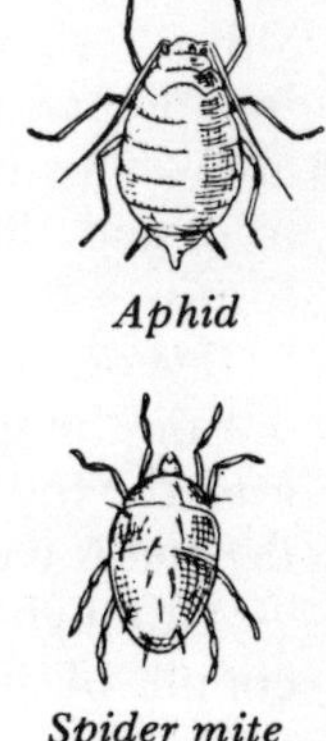

Aphid

Spider mite

Spider mites. The tiny red spider mites are also common, with their fine webs. They are especially fond of backyard gardens, feeding on beans, tomatoes, eggplant, onions, celery, and melons, in particular.

Beetles. If you are growing cucumbers or other vine crops such as squash, your special problem will be cucumber beetles. They are greenish or greenish-yellow in color, banded or striped.

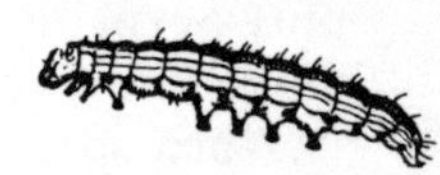

Armyworm

Armyworms. Many vegetables are attacked by armyworms, so-called because they maneuver in large groups. They are up to two inches long, light tan to dark green or black caterpillars with white stripes along the sides and back.

Thrip

Thrips. Onions and many other vegetables are affected by onion thrips that have dark bodies not over a sixteenth of an inch long. Other species of thrips, usually brownish or yellow, favor asparagus and beans.

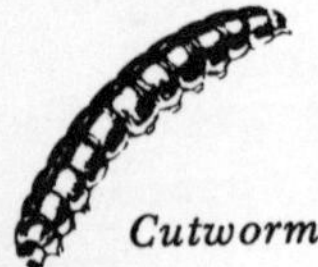
Cutworm

Cutworms. Then there are the cutworms—dull-gray, brown or black caterpillars, the larvae of moths. Vegetables are especially susceptible to black cutworms if the crops are grown in rows or on hills. Other species are particularly destructive to early-season plantings or a variety of vegetables—peppers, tomatoes, cabbage, peas, beans, beets, and squash.

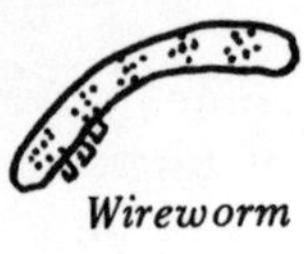
Wireworm

Wireworms. Finally, there are various species of wireworms, the larvae of click beetles. The insects are intriguing—when they aren't eating plants. When upset, they can right themselves, with a quick clicking flip into the air they land on their feet. They are hard on onions, carrots, beets, corn, beans, lettuce, turnips, and potatoes.

Both cutworms and wireworms are best controlled by soil treatments, rather than by spraying plant foliage as is usually done. This can be accomplished either with a granular material or liquid. Spectracide and Sevin, for example, come in both forms.

If you use a granular insecticide, either before planting or during the growing season, make sure it gets thoroughly worked into the top inches of soil. If you spray the chemical on, be sure the soil is well saturated.

Although the list of bugs that like vegetables seems tiring, a couple of sprays available at every garden shop will get rid of them. General purpose sprays contain Malathion or Diazinon plus Sevin or methoxychlor. Read the containers carefully to make sure you don't apply them too close to harvest.

When spraying, cover both sides of the plant foliage until you get a light drip. It's especially important to reach hard-to-get-at undersides of leaves and stems because that's where insects often hide.

By planting not just enough for a family but also for the bugs, many gardeners find frequent spraying isn't as necessary and the cost of growing the extra plants is minimal, offsetting the cost of pesticide treatment.

One look at everything that can go wrong in a vegetable patch and the sensible gardener is likely to throw in the trowel.

Not so fast. The entire bug world is not going to descend on your plot, nor will blights in all colors. But if you notice something gnawing on your cabbages, it's nice to be able to chastise the creature by name.

If you were to determine the vegetables you should avoid growing by the possible number of problems they might have, certainly everyone would avoid tomatoes, yet they are the most widely and successfully grown vegetable in South Florida. Especially if you are on a regular spraying routine with a combination insecticide-fungicide, you may never become acquainted with more than a few wayward insects.

Here are the pests and conditions that can crop up. Some novices are tempted to buy a different spray for every problem, but it pays, in both time and money, to call around and locate combination sprays. One of these combinations, supplemented by, perhaps, one or two other materials for particular conditions, makes a picture-perfect harvest easy.

For photographs of many of the insects named, refer to Lewis Maxwell's booklet "Florida Insects" (see p. 123).

Beets. A cool-weather vegetable that likes a very moist soil for good germination. Nematodes, caterpillars, and fungus diseases can happen to them.

Mistakes beginners make: Not eating their leaves and stalks, as well as their root. Also, stringy, tough beets are the result of insufficient moisture or competition from weeds or other beets. Beets must be thinned adequately to prevent a tangle of roots.

Broccoli. A shallow-rooted crop that needs good moisture. At its best when it matures in cool weather. It can be harvested over a long period—first the head and then the many side shoots that continue to develop. Don't let flowers open.

Broccoli can be hit by several diseases (downy mildew among them) best prevented by regular spraying. Caterpillars and plant-bugs can be a nuisance.

Brussels sprouts. A reluctant yielder in South Florida, it likes the coldest weather best and is hurt by several days of warm weather midwinter. Problems could be diseases, including downy mildew (spray preventatively), aphids, caterpillars, and plant-bugs.

Cabbage. A gross feeder that likes cold weather. Grow in plot from transplants raised in flats, and plant deeper than they were growing. Watch for imported cabbageworm. Cabbage looper, a major pest, is hard to control as it matures but when young, a weekly preventative spray program will get it. On maturing worms use Thuricide, Dipel, or Biotrol.

Cabbage aphids are seen on several crops, including turnips. Prevent the population from building up so leaves won't become distorted, providing a hiding place from sprays.

The harlequin cabbage bug is another familiar pest on this vegetable. Stunting and wilting of leaves are symptoms of the several plant bugs that damage vegetables. They can kill a plant and are hard to control, if not impossible, when mature.

Black rot is identified with cabbage and it is a good idea to secure resistance varieties. Maxwell recommends weekly spraying with neutral copper to protect plants not yet infected. This disease causes light brown or yellow areas with a network of black veins along the edges of the leaves, in V-shaped areas.

Alternaria leaf spot may attack the curds of cauliflowers, the older leaves of cabbage and broccoli—and the leaves of turnips and radishes. Small dark areas enlarge and spread during damp weather, showing concentric rings. Linear spots appear on the

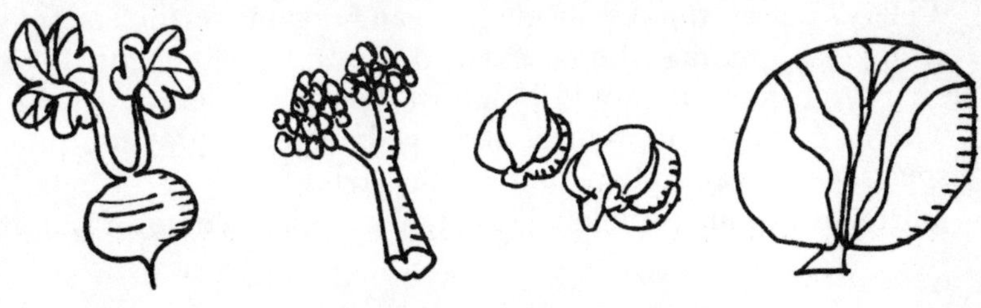

stems and leaf petioles. Weekly spraying with a fungicide is effective.

Downy mildew plagues cabbage, cauliflower, collards, Chinese cabbage, Brussels sprouts, broccoli, kale, and kohlrabi during cool, wet weather. It is very hard on seedling plants. It first appears as a white to lavender fluffy mold on the undersides of leaves. Scattered pale spots appear next on the head of cabbage and leaves of the other crucifers (cabbage relatives mentioned before). As with many fungus related problems, secondary organisms can enter plants through these lesions, so fungus should be kept at bay with regular spraying.

Cantaloupes (including honeydews and casabas, etc.). Honeydews are rarely up to par in South Florida and cantaloupes prefer a drier climate on the warm side. So try these melons in the late winter to early spring. Cantaloupes are also heavy feeders.

Grow only varieties recommended for South Florida, and fend off downy mildew with regular fungicide spraying. Gummy stem blight is another cantaloupe disease.

Carrots. They do best only during the coolest months in a rich deep soil. Foliage diseases cause less damage than nematodes, but they are prone to some.

Mistakes beginners make: Not thinning out plants adequately. To reach full size they need room and little competition. Immature carrots can be eaten in salads.

Cauliflower. This vegetable isn't as easy to grow as its sister, broccoli, but shares its problems. It is not quite as hardy as another relative, cabbage. Attracted to it are caterpillars, plantbugs, and several diseases that call for regular fungicide use.

Mistakes beginners make: To make the head curd (turn its whitest), you may have to cover the head when it is about two to three inches across. Pull its leaves over it and tie them together. This will bleach the head.

Celery. Celery thrives on manures and organic fertilizers; several applications may be necessary for best development. Like all hardy growers, it should be grown during the coolest weeks.

Celery is prone to fungus and needs regular spraying, plus inspections for caterpillars that eat the stalks.

Mistakes beginners make: This vegetable likes plenty of water

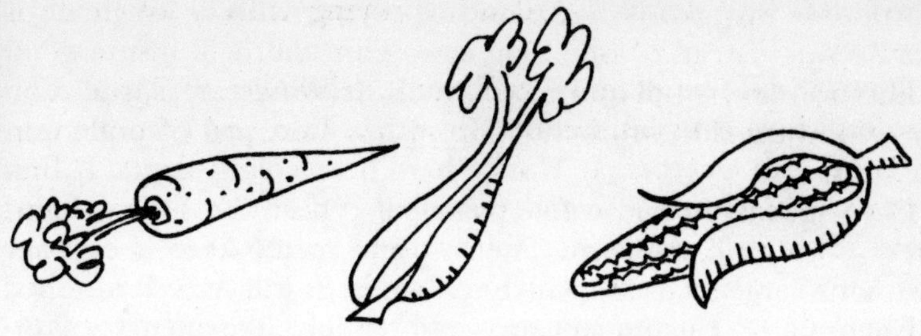

and good air circulation. Unless the weather stays cool and the soil moisture remains high, it won't perform.

Chinese cabbage. Another coolest-weather grower, this plant makes good salad and cooked greens and is bothered by little more than leaf-eating caterpillars. It needs soil kept consistently moist. It should be picked if a heat wave is expected since mild weather will make it bolt (flower) and it will become tough.

Collards. A cabbage that does not form a head, collards thrive during the winter months, especially after a frost. But they can be grown during the summer months with care. Easier to grow than cabbage, they need attention to fertilizing to be succulent.

Caterpillars, plant-bugs, and crucifer diseases (controllable by regular spraying) are possible problems.

Sweet corn. Corn is most famous, problem-wise, for the corn earworm. To stop this pest, you should spray with Sevin or a material containing Sevin every other day from the time the first silk appears until the last one turns black. Corn budworm is best controlled by putting a small pinch of cutworm bait in each bud of small corn plants. It is also cold-tender.

Northern and southern leaf blights come during wet weather. They begin as spots and elongate to tan areas about four inches long on the leaves. If it is a particularly rainy season, spray with a fungicide twice a week.

Bacterial leaf stripe begins as water-soaked, olive-green stripes and blothes running parallel with the leaf veins and becoming yellow or tan when the tissue dies. It starts with lower leaves. There is no practical control.

Mistakes beginners make: Wrong planting time. Corn grows best in a mean temperature range of 60 to 75 degrees. (Monthly maximums should not average over 95 or below 50 degrees.)

Another mistake is not planting corn in blocks of three or more rows instead of one long row. This will help insure good pollination and a full quota of kernels. In windy areas, stalks on the windward side will benefit from hand dusting of pollen on the silks.

Corn is classified into five principal types: dent, flint, flour, sweet corn, and popcorn. The soft and mealy kernels of flour corn with large seeds are widely grown in South America.

Cucumbers. Cucumbers are cold-tender, like plenty of organic fertilizer, and need a constant level of moisture. They should be grown on nematode-treated or nematode-free soil.

Pickleworms, which bore into the fruit, like this vegetable. Melonworms favor the foliage but may go into fruit. If you want to be sure to keep these pests away, spray twice weekly with Dipel, Thuricide, or Biotrol, covering all of the plant. Sevin or rotenone may be used, but spray late in the day. Don't spray blossoms as you don't want to kill insects valuable for pollination. Cucumbers and watermelons (other vegetables, also) are affected by viruses on occasion. The cucumber mosaic virus extends the blotched and spotted look implied by its name to both leaves and fruit. The best way to avoid viruses is to select resistant varieties and to destroy all virus-infected plants in the vicinity.

Powdery mildew is very common on cucumbers in cool, damp weather. It begins on the undersides of older leaves as small white spots and spreads until the leaves look powdered, then turn brown. Young stems are also killed. Fruits on infected vines are stunted and ripen prematurely. Karathane or Benomyl should be used before it appears in a regular spray program—twice weekly during cool, wet weather.

Downy mildew is occasionally seen on cucumber leaves and stems. It is probably the most serious disease of cucurbits (cantaloupes, watermelon, squash, and related vegetables). It starts on the upper surfaces of leaves as light areas. Underneath, on the undersides of the leaves, are the downy gray areas that later become brown and angular. Spray with fungicides other than Karathane or Benomyl-Benlate, and destroy affected leaves.

Benomyl (also zineb, Maneb, Dithane M-45, and Bravo) is effective against anthracnose on cucumber leaves and stems. This problem appears first as small, brown, water-soaked spots

Thriving on organic fertilizers, cucumbers need a steady moisture level and relatively nematode-free soil. They are famous for mildew problems that can be controlled with early spraying.

that enlarge quickly and turn brittle. Then the lesions fall out. This disease shows as round water-soaked areas on fruit. They eventually turn brown. In wet weather they are rather pinkish.

Mistakes beginners make: Letting fruit mature on the vine. This will make the vine stop producing. Young plants are seriously injured by frost; most favorable daily temperature is 65-75 degrees.

Eggplant. Eggplants have a long life—perhaps three years or more—if they escape cold damage. They are gross feeders and like to be kept uniformly moist. Avoid droughts. Plant them in nematode-free soil, and keep red spider mites away with Kelthane. For pinworm (it makes a large blotch mine on leaves) you may find Cygon better than the Diazinon used in many combination sprays. Or vise versa. Opinions differ. Repeated spraying may be necessary.

Mistakes beginners make: Planting too early. Low night temperatures cause blossom drop.

Endive and escarole. Easy growing. Watch for signs of bolting (flowering) and pick before it happens. (See lettuce.)

Kale. Caterpillars, plant-bugs, and diseases should be kept under control with a combination spray.

Kohlrabi. Easy-growing cold-lover. Watch for caterpillars.

Mistakes beginners make: Waiting too long to pick. Cut leaves under three inches in diameter, over an inch and a half.

Lettuce. Likes stable growing conditions of cool weather and even soil moisture (no drought). Thrives on organic matter. A warm spell encourages bolting after which quality deteriorates. Bibb lettuce is easier to grow than head lettuce; Cos or Romaine is the easiest of all. Leaf lettuce stands warm weather better.

Watch for aphids, slugs, and snails (use beer or metaldehyde bait). Lettuce drop is a fungus that prompts older leaves to wilt, inner leaves to become soft, then dry and dark in color. A white growth shows around the main stem. Soil fumigation helps pre-

This leaf lettuce is growing in plastic bags of sterilized potting soil with tin cans inserted for easy watering. Even soil moisture, along with cool weather, is important for a good crop.

vent this. If infection shows, destroying the plant is all that can be done—and not replanting in the same spot with lettuce.

Leaf mustard. Care resembles that of cabbage. Mustard is easy to grow. Caterpillars, plant bugs, and several diseases should be warded off with regular combination spraying.

Lima beans. Lima or butter beans are gross feeders (they like rich soil), they prefer warmer and more humid weather than snap beans, and are occasional hosts to plant-bugs. (See snap beans for related problems.)

Okra. Gross feeders that like moist soil. Also warm weather, nematode-treated soil. Watch for stink bugs.

Mistakes beginners make: Picking and using mature pods that are too woody. Immature pods should be harvested daily.

Onions. Keep moisture in soil at a consistant level or plants will not recover fully. Plant recommended varieties for few problems.

Mistakes beginners make: Planting sets too deeply. Put just below soil.

English (garden) peas. Cold weather is a must. Dwarf varieties grow best. Preventative sprays are necessary to hold back leaf diseases.

Mistakes beginners make: Picking peas after they become mature and hard.

Southern peas (cowpeas). They prefer very warm weather, a low nitrogen fertilizer, and a soil that is not too wet. Problems include nematode-sensitivity, cowpea curculio, caterpillars, and plant-bugs. Use a combination spray regularly.

Bell or sweet pepper. Organic fertilizers are appreciated. Consistent moisture level too. Problems could include nematodes, fungus, bacterial, and virus disease, and thrips that can cause young fruit and flowers to drop. Bacterial spot makes dark brown translucent spots on the leaves and fruit, causing them to drop if the infestation is bad (as it can get in wet, cool, windy weather). A weekly spraying with neutral copper combined with Maneb or Dithane M-45 from the time the plants are set out will help control this.

This spray schedule will also take care of frogeye spot on leaves, a condition that shows as spots with light tan centers. A heavy infestation will cause severe leaf drop and a reduction in the number of peppers.

Sevin or Diazinon in a combination spray will handle thrips that can damage the blossoms on bell peppers and cause small peppers to drop.

Potatoes. Red potatoes do best in South Florida. Seed potatoes are cut so that each eye (or more than one) is on a cube at least an inch and a half square. The soil should be nematode-fumigated. A combination spray will control blight, caterpillars, and the potato beetle, also affected by rotenone. A menagerie of diseases have been reported on potatoes, so weekly preventative spraying and the purchase of recommended varieites should be followed.

Mistakes beginners make: Overfeeding with nitrogen fertilizer.

Sweet potato. Cold-tender plants, the red sweet potato varieties are not real yams but are native to the tropical Americas. As with potatoes, don't overdo organic fertilizer. The sweet potato weevil and wireworms should be controlled by preventative spraying. Same with rot and wilt fungi. Use plants that have clean roots.

Snap beans. These are immature bush, pole, or wax podded beans. The most common bean disease is bean leaf rust, a fungus that attacks leaves and sometimes pods. It shows on the upper sides of leaves as brown dots in the middle of pale yellow spots. Eventually the whole leaf will turn yellow. Some varieties are more resistant than others. This condition can be met best before it occurs with a preventative spray program using a fungicide. It is most common in warmer weather.

Another warm weather (early fall and late spring) disease is white mold (Southern blight), characterized by wilting and a white collar at the base. It is most common on old bean-planted soils. There is no control.

A virus disease that sometimes shows up is common bean mosaic. A downward curling of leaf margins and a puckering is characteristic. This problem stunts plants and reduces yields. It is transmitted mainly by aphids. Virus-resistant varieties are available. Aphid control on nearby plants helps. Bean pests include bean leafhoppers, bean leafrollers, and Mexican bean beetles.

Mistakes beginners make: Planting seed too deep and waiting to pick after they've passed the prime young succulent stage.

Radish. So easy a child can grow it well. Root maggots are seldom a problem.

New Zealand (summer) spinach. An easy-to-grow, warm weather plant. Caterpillars are the only significant pests.

Mistake beginners make: Not nicking large seeds with a knife to speed up slow germination.

Summer squash. It can also be a fall or spring squash. Summer squashes include zucchini, crooknecks, and straightnecks. Hard to transplant and cold-tender, they are usually sown directly in the garden.

Bugs that like squash include pumpkin bugs, pickleworms, and caterpillars. Pickleworms bore into fruit while melonworms largely prefer foliage. Both can be controlled with Sevin or rotenone used twice weekly, late in the day and not on blossoms. Dipel, Thuricide, or Biotrol are other controls that can and should cover flowers, fruit, and growing tips. They will not hurt pollinating insects.

Pickleworms can sometimes be trapped by placing boards on the soil around the plants. Bugs will hide under the boards and the bugs can be found there and removed in the mornings. Sevin or Diazinon will also control beetles.

Squash vine borers can be controlled by splitting one side of the stem where they have entered with a razor blade and puncturing the worms. Then put a mound of moist soil around each cut stem to prevent drying and to induce root growth beyond the point of injury. Spraying the vines with Sevin once a week helps, starting when runners develop.

Blossom blight (wet-rot) sometimes appears on young fruit as a fuzzy black mass. Fruits turn yellow and rot. A weekly spraying with zineb helps curtail this.

Powdery mildew is a common squash problem, spotting the undersides of older leaves, then spreading to cover the upper sides with a sugary powder. Leaves die, starting with the older ones, and fruit is stunted. Karathane and Benomyl give good preventative control; a regularly used combination spray will also help.

Downy mildew starts out as light areas on the leaf tops, downy gray underneath. Spots later become brown and angular. A weekly spraying with zineb, Maneb, Dithane M-45, or Bravo is the control. Destroy affected plants.

Mottled green areas and lumps on fruit indicate the presence of mosaic virus. The only control for this is preventative; don't let your garden harbor aphids and beetles that transmit the disease.

Mistakes beginners make: Underrating the importance of regular preventative spraying on squash. Also, allowing fruit to get oversized. Important: see section on pollination of vegetables, p. 77.

Winter (running) squash. These are the hard skinned squashes—Butternut, Acorn, and Buttercup. Picked immature, they resemble summer squash in taste. They need room to grow. They have the same potential problems as summer squash.

Strawberries. A special care plant. Plantings are made in October and February. Best beds are raised—four to six inches high, three feet wide, with two rows of plants. Soil is fumigated for protection against nematodes, a serious problem on strawberries. Fertilizer is mixed in at rate of seven pounds per 100 square feet. Cover beds with black plastic held down with rocks or bricks. Make openings in plastic a foot apart.

After planting transplants, water them well with a liquid 20-20-20 fertilizer. Subsequent waterings should be frequent enough so plants never wilt.

The wireworms (click beetle larvae) and caterpillars (larvae of cornstalk borers, generally) that can attack strawberry fruit are not as prevalent on plants grown in plastic, but, if you spot them they can be controlled with Diazinon. Since Captan, zineb, and Benlate are the only three fungicides approved by the Environmental Protection Agency for use on strawberries, a combination spray cannot be recommended unless it contains only Captan. (Other fungicides may not be harmful but they have not been put through the rigorous testing process yet for EPA approval.)

To avoid the small purple spots with red centers that are called common leaf spot on strawberries, you should spray every five to seven days with one of the fungicides mentioned above. They will also help control pestlotia fruit rot and anthracnose fruit rot.

Spider mites—those tiny, almost invisible bugs—can be devastating on strawberries. They give leaves a dry, dusty appearance,

Here is an unusual way to grow strawberries—in "towers" made from poultry wire and lined with plastic. This gardener also caged his plants, determined to keep birds away.

eventually turning them brown. Mites can become resistant to a single chemical used regularly, so alternating with Kelthane and Malathion on a weekly schedule is recommended.

Swiss chard. A cool weather grower, Swiss chard should have its leaves removed from the outside to prolong the life of each plant. Its fleshy stems are eaten, too. Caterpillars are the most significant problem if they appear. Use a combination spray.

Tomatoes. Most popular and one of the plants most plagued by bugs and diseases. Despite this, many gardeners grow them with ease. Tomatoes must have full sun for good production. Moisture in soil should be kept constant, rather than varying between drought and deep soaking. A nematode-free soil is very important, but just as significant to success is the selection of recommended varieties for South Florida that are especially disease resistant.

Keep in mind that high-nitrogen fertilizers or organics will produce plenty of vines but won't contribute much to fruit set. (See section on *Encouraging Fruit Set.*) A good 6-6-6 will keep plant stable after it is mature. (Extra nitrogen will help little plants get going.)

The use of a combination insecticide-fungicide spray every five to seven days makes growing tomatoes much simpler than trying to attack each setback as it appears.

Leaf miner is probably the first visible problem most gardeners encounter. It can cover leaves with winding tunnels that eventually cause them to dry out and turn brown. Diazinon is its best control. It is almost impossible to get rid of miners entirely.

Tomato pinworm is a villain on fruit, spoiling it with its tunnel, usually under the calyx or into the shoulder. It also can make a broad, rounded mine in the leaf, turning it brown. Preventative spraying is the best control. Diazinon will also take care of the pinworm.

The southern armyworm is a common brown, black, and gray banded worm that eats leaves and fruit and can be controlled with a spray containing Diazinon, Sevin, Dipel, or Thuricide.

Tomato hornworms are usually hand picked and disposed of. These large caterpillars are very hard to kill with sprays.

Of the fungi that like tomatoes, late and early blights are most commonly encountered. (Their names originated in the northern growing areas and are not too relevant here.)

Late blight is the same disease that was responsible for Ireland's Great Potato Famine. It appears mostly from November through February when nights are cool and moist (temperatures from 50 to 70 degrees F. at night, 70 to 80 degrees F. in the daytime) with high humidity.

It affects leaves, stems, and fruit, leaving large, irregular, water-soaked areas on the leaves that eventually will turn them brown. Under dew conditions, a fine white mold can form along the edge of the affected areas.

On fruit, the lesions are large, mahogany colored, and water

The tomato is South Florida's most popular vegetable, as it is nationwide. Varieties popular in other parts of the country often do poorly here, however. Pick one of those recommended for South Florida.

soaked, usually on the upper part of the fruit. Stem lesions can girdle and kill the plant.

Early blight's prime temperature is around 75 degrees F., during rainy weather. The dark lesions usually contain concentric rings. Areas around the leaf lesions will turn yellow and eventually cause the leaves to drop. Stems are often girdled, and the plant is killed.

Gray leaf spot is a leaf-only disease that covers the plant with tiny spots. It is best avoided by planting resistant varieties (see Planting Guide for Vegetable Gardens chart in Appendix.)

Fusarium wilt is a common soil-borne disease affecting only one side of a leaf or plant. Use only resistant varieties; spraying cannot control it.

Daconil (Bravo) is favored by many as a fungicide for all of these problems on tomatoes. Also good are zineb and Maneb. An every five-to-seven-days spraying schedule is very important on tomatoes. When weather is wet and humid or wet and cold, spraying can be done every three days for good measure.

Bacterial spot sometimes is found on tomatoes, particularly hydroponic tomatoes. Neutral copper or combinations of neutral copper and Maneb or neutral copper and Dithane M-45 (Fore) can be used as a preventative, but bacterial diseases are harder to control than fungus problems.

Mistakes beginners make: Buying seedling tomatoes that are too large. Buying varieties other than those on recommended list. Planting in semishade. Not spraying until problems show up. Planting in untreated garden soil. Letting soil dry out thoroughly between waterings.

Turnips. Eating tops as well as roots extends the usefulness of this small plant. Watch for aphids and leaf spot diseases. Keep soil moist.

Mistakes beginners make: Turnips are a short-term crop and need frequent light fertilizing for best greens.

Watermelons. They take up a lot of room but few people can resist trying to grow them. Watermelons need a rich, nematode-treated soil and should be planted when all danger of frost is past. Several fungus diseases, including gummy stem blight, anthracnose, and downy mildew, choose watermelons, as do some virus diseases. Aphids and rindworms are their principal bugs.

An every-five-to-seven-days spray program is almost mandatory in South Florida. Choose a good insecticide-fungicide combination and consult the individual sprays that are used on squash diseases if further control is necessary.

Mistakes beginners make: Not enough watering. Not only are these plants the largest water users but the extra warm South Florida sun in the spring and summer removes water from the ground very rapidly.

How Harmful Are Pesticides?

All pesticides are poisonous to humans when misused. When used with care, they make possible a healthy, inexpensive harvest that most busy people could not otherwise enjoy.

Pesticides should be applied selectively and carefully. Do not use them when there is any danger of drift to other areas or onto the person spraying. Avoid prolonged inhalation of a pesticide, spray, or dust. It's a good idea to wear long sleeves and trousers when you use them.

After handling a pesticide do not eat, drink, or smoke until you have washed. In case a pesticide is swallowed and gets in the eyes, follow the first-aid treatment on the label and get prompt medical attention. If a pesticide is spilled on your skin or clothing, remove clothing immediately and wash skin thoroughly. Then launder the clothing before wearing it again.

Chlordane, Diazinon, Dimethoate, Naled, and toxaphene can be absorbed directly through the skin in harmful quantities. When working with them in any form, take extra care not to let them come in contact with the skin.

Be sure to apply pesticides only to those crops for which it is recommended. Allow a sufficient waiting period—at least one day—before a harvest, or longer if the label specifies. Then wash all treated vegetables before eating.

Carbaryl. Do not apply Carbaryl to asparagus, beans, carrots, cucumbers, eggplants, melons, okra, peas, peppers, pumpkins, squash, sweet corn, or tomatoes within 1 day before a harvest; or to broccoli, brussels sprouts, cabbage, cauliflower, head lettuce, kohlrabi, parsnips, radishes, or rutabagas within 3 days before a harvest. Do not apply Carbaryl to blackberries or raspberries within 7 days; or chard, Chinese cabbage, collards, kale, leaf lettuce, mustard, or spinach within 14 days. Do not apply Carbaryl to table beets or turnips within 14 days before a harvest (3 days if tops won't be used for food or feed). Do not apply Carbaryl to onions or potatoes.

Chlordane. Do not apply chlordane to any plant after the appearance of foliage or fruit that is to be eaten or fed to livestock. Do not repeat soil application of chlordane for at least 3 years. Do not apply chlordane to soil within 2 years before planting asparagus, carrots, parsnips, or radishes.

Diazinon. Do not apply Diazinon to peas or tomatoes within 1 day before a harvest; or to melons or winter squash within 3 days; or to broccoli, cauliflower, or peppers within 5 days; or to beans, blackberries, brussels sprouts, cabbage, cucumbers, raspberries, or summer squash within 7 days; or to carrots, celery, collards, kale, lettuce, onions, parsnips, radishes, spinach, or turnips within 10 days; or to table beets within 14 days; or to potatoes within 35 days before a harvest. Do not apply Diazinon to foliage of asparagus, eggplants, kohlrabi, okra, pumpkins, mustard, or rutabagas.

Dicofol. Do not apply Dicofol to blackberries, cucumbers, melons, peppers, squash, raspberries, or tomatoes within 2 days, or to beans within 7 days before a harvest. Do not use Dicofol on potatoes.

Dimethoate. Do not apply Dimethoate to peppers within 1 day before a harvest, to tomatoes within 7 days, or to spinach within 14 days.

Endosulfan. Do not apply Endosulfan to cucumbers, eggplants, peppers, potatoes, pumpkins, squash, or tomatoes within 1 day before a harvest; or to broccoli or cabbage within 7 days; or to brussels sprouts or cauliflower within 14 days; or to collards, kale, mustard, spinach, or turnips. Do not feed treated plants to livestock. Do not apply to beets, beans, kohlrabi, or chard. Apply only to plants for which it is recommended on the label.

Fungicides. Do not apply zineb to beets or carrots within 7 days before a harvest if the tops are to be used as food. Do not apply ferbam or ziram to carrots within 7 days before a harvest if the tops are to be used as food. Do not apply zineb to chard or spinach within 7 days before a harvest. Do not apply ferbam to blackberries or raspberries within 40 days. Do not apply ziram on potatoes. Do not apply ferbam on potatoes or sweet potatoes. Fungicides may be used up to 1 day before a harvest on other crops for which they are recommended.

Malathion. Do no apply malathion to asparagus, beans, blackberries, cucumbers, melons, okra, squash, tomatoes, or raspberries within 1 day before a harvest; or to broccoli, eggplants, onions, peas, peppers, pumpkins rutabagas, or turnips (including tops) within 3 days; or to sweet corn within 5 days; or to beet tops, brussels sprouts, cabbage, carrots, cauliflower, celery, collards, kale, kohlrabi, head lettuce, mustard, radishes, or spinach within 14 days before a harvest.

Methoxychlor. Do not apply methoxychlor to cantaloupes, cucumbers, eggplants, kohlrabi, peppers, pumpkins, squash, tomatoes, or turnips (if tops are not to be used) within 1 day before a harvest; or to asparagus, beans, blackberries, black-eyed peas, cabbage, or raspberries, within 3 days; or to beets or carrots (if tops are not to be used), cauliflower, peas, radishes, rutabagas, or sweet corn within 7 days; or to beet tops, broccoli, brussels sprouts, carrot tops, collards, kale, lettuce, spinach, or turnip tops within 14 days before a harvest. Do not apply methoxychlor to onions or okra.

Naled. Do not apply Naled to broccoli, brussels sprouts, cabbage, cauliflower, chard, cucumbers, eggplants, lettuce, melons, peppers, pumpkins, spinach, squash, tomatoes, or turnips within 1 day before a harvest. Do not apply Naled to beans, table beets, blackberries, celery, collards, kale, kohlrabi, mustard, okra, onions, peas, potatoes, raspberries or sweet corn.

Toxaphene. Do not apply toxaphene to tomatoes within 3 days or to eggplants or peppers within 5 days before a harvest. Do not apply to celery after bunch begins to form or stalk is half grown. Do not apply to other crops after appearance of parts to be eaten or fed to livestock. Do not apply toxaphene to asparagus, beets, chard, endive, melons, mustard, potatoes, squash, sweet potatoes, or turnips.

Dispose of empty pesticide containers by wrapping them in several layers of newspaper and placing them in your trash can.

It is difficult to remove all traces of a herbicide (weed killer) from equipment. Therefore, to prevent injury to desirable plants do not use the same equipment for insecticides and fungicides that you use for a herbicide.

Stress

A plant can be under stress for several reasons, all of which weaken it so that when an additional adverse condition comes along—like an infestation of pests or cold weather—the plant has no resources to withstand the problem.

The most common cause of stress is uneven care—no water for some time, followed by excessive watering. Overfertilizing, an unchecked insect infestation or disease problem, a cold snap, and hot weather are all possible points of stress.

Some vegetables are more prone to suffer from stress than others. Lettuce and eggplants will taste bitter, cantaloupes, turnips, and mustard greens not as sweet, radishes will be pithy.

Many gardeners have had the experience of slicing open a fresh, good-looking cucumber only to find the flesh so bitter that it could not be eaten. Insufficient moisture—a marked variation in soil moisture—is believed to be one cause. Some growers feel that bitterness is more prevalent during cool growing seasons than during warm ones. Faulty fertilizing and even harvesting during the wrong part of the day are other theories advanced when bitterness crops up.

Encouraging Fruit Set

Fruit set problems most often occur with tomatoes and the cucurbits—squash, cucumbers, melons.

Cucurbits bear male and female flowers on the same plant. The female flower must be pollinated before the ovary or fruit will develop. The male flower differs from the female in that the ovary appears between the flower and the stem petal on the female flower, but the male flower is directly connected to the flower stem.

The female should be quickly apparent since the ovary stands out as a miniature fruit—so much so that some gardeners mistake it for a developing, pollinated fruit.

Usually the curcurbits' male flowers open before the female flowers do, which makes pollination rare unless there are plenty of bees in the neighborhood.

If bees are in short supply, you can hand-pollinate the female flowers. Select an open male flower and pick it from the plant. Remove the petals to expose the stamen (the long "feelers" tipped by the pollen-covered anthers). Gently rub the pollen-bearing anther across the stigma of the female flower.

If many male flowers are available, deposit pollen from several male flowers on the stigma of each female flower. Pollination should be carried out each morning before 9:30 a.m. as pollen is ripe early in the day and dies before noon. New flowers (both male and female) open every day. The ovary will wither, rot, or, if it remains on the stem, not produce a normal fruit if the flower is not pollinated.

While many gardeners like to mist tomato blossoms with an aerosol can of hormone bloom set, fruit-setting reluctance on these plants often goes beyond this remedy.

Failure of fruit to set on tomatoes is the most common problem among gardeners, especially in the spring. Plants can be large and healthy, covered with yellow flowers, yet the blossoms mysteriously drop off or wither instead of producing fruit.

Here are some causes:

High nighttime temperatures. Fruit set in tomatoes is very dependent on temperature. The varieties most desirable for planting in South Florida will set fruit when nighttime temperatures are about 70 degrees F. There is practically no fruit set when night temperatures are 90 degrees F. in the soil.

The set of tomato fruit is determined long before the bloom appears. The first two clusters are affected when the first and second leaves are only an inch and a half long. Other clusters follow the same pattern.

Low nighttime temperatures. During extended periods where the temperature at night drops below 55-58 degrees F., tomato varieties recommended for Florida will not set fruit abundantly.

Excessive nitrogen. Tomato plants tend to initiate flower buds under normal conditions as soon as they have reached a certain size or age. This is the time when the plant changes over from a strictly vegetative form of growth to a reproductive one. Generally, nitrogen applied in excessive amounts before flowering will cause the plant to continue vigorous vegetative growth rather than setting fruit. This is frequently the case with tomato plants growing over septic drains, in rich organic soils, or in heavily manured beds.

Tomato production usually is best when only enough nitrogen is available during the first couple of months after planting to permit development of a large sturdy vine—but not enough to cause a soft type of growth. Once the plant has a heavy fruit load and tomatoes begin to ripen, the nitrogen requirement is again very high.

If the nitrogen supply is not adequate at this stage of development, the plants will turn yellow and defoliate, leaving the fruits exposed and vulnerable to sunscald.

Excessive shade. When plants are not especially vigorous or dark green in appearance, it may be that they are in too much shade—which also reduces fruit production.

Drought. Water stress often keeps flowers from producing a mature fruit. It may be either from interference with the flower fertilization or a disruption in the growth of the little fruit following fertilization.

Insects. Various insects, both sucking and chewing, can feed on the ovules, flowers, and partially developed seeds and fruits. Blossom or young fruit drop may result. The culprits may be thrips, pinworms, grasshoppers, aphids, or spider mites.

Cold Weather Protection

Gardeners should keep in touch with weather reports during the winter, and if there is any question about the temperatures going into the 30's, they should take precautions. Even in the 40's, if strong winds are expected, protection should be given to a small plot.

When there is no wind, sprinklers can be turned on and left to run until morning. If there is wind, sprinklers should not be used as frost can form on vegetables not kept consistently watered.

Simply watering the plot well before a cold front arrives is a form of protection since the evaporation will raise temperatures somewhat. Covering the plot after watering affords the best protection if it is practical. Large sheets of plastic can be used, or even newspaper watered down and anchored with rocks. Coverings should be removed as soon as possible in the morning.

Some forms of cold damage–on leaves, for instance–show up immediately and others–such as fruit formation and development–may take longer. All badly affected plants and plant parts should be removed so that fungus problems are not allowed to set in and damage the rest of the crop. Fungicide spraying should be increased to every third day when plants are subjected to extra watering and covering in cold weather.

Cucurbits (squash, cucumbers, melons, etc.) can really be hurt by frost. Other vegetables that can't take the cold are tomatoes, young potatoes, sweet and field corn (especially if in silk and just starting to tassel) beans, eggplants, okra, peppers, and sweet potatoes. Not as susceptible are cabbage, broccoli, cauliflower, collard and mustard greens, and onions.

Damage sometimes does not show up until three weeks after a cold snap. In the case of vegetables like tomatoes or beans, the cold may ruin the current crop but the plants will continue to grow.

When to Harvest Your Crop

One of the big advantages of growing vegetables in your own garden is you can harvest them when they are at their best. It can make a big difference in the way they taste.

Lettuce is without a doubt the world's most popular salad plant. Pick looseleaf lettuce as soon as the leaves are large enough to use, and the others as soon as they have headed. While firm heads are desirable for many uses, the very hard head is overmature. Such heads may have discoloration in the midribs and a bitter flavor. Lettuce becomes bitter after the seed stem appears.

Cold, clean, and covered refrigerator storage is a must for all green leafy vegetables. Lettuce needs the additional precaution of separate storage. A gas given off during the ripening process of many fruits may cause lettuce to develop brown spots throughout the head. If you can't provide a separate crisper, seal the washed and dried head in a plastic bag in the refrigerator where it should keep its good quality for at least a week.

Tomatoes in the garden should be harvested when they are fully vine ripened. The quality is usually highest when color is a dark red (in red varieties).

The red pigment in tomatoes does not form when temperatures are above 86 degrees. That is why tomatoes are yellowish-orange instead of red when temperatures are high. If the plant has lots of foliage to shade the fruit, the tomatoes will color better.

Tomatoes should not be put into the refrigerator until they are fully ripe. If you get tomatoes that are not red-ripe, keep them at room temperature away from direct sunlight. A windowsill is not a good place, because sunlight may prevent even ripening and cause the tomatoes to wither and soften. Tomatoes actually will ripen well in the dark.

To peel or not to peel your tomatoes? It is a matter of preference. If you want to peel them, here is how: Hold the tomato on a slotted spoon and dip into boiling water for about a minute, then into cold water and slip off the skin.

Sweet corn is at its best when the silks start to turn black. The kernels should then be plump, bright, and full of milk. The quality of sweet corn depends on the amount of sugar in the kernels. If corn is not cooled quickly and kept under refrigeration (at about 40 degrees) after being picked, the sugar rapidly turns to starch and the sweet taste is lost. On a warm day, sweet corn left unrefrigerated will lose up to 15 percent of its sugar content in 3 hours, and up to 50 percent in 24 hours.

If you don't have room enough in your garden to grow sweet corn, find a roadside stand that sells it and arrange to get there immediately after the corn has been picked. Get it home and into the refrigerator as soon as possible. Kept in the refrigerator it should stay sweet two or three days. It helps to take a container with ice in it out to the roadside stand and put the corn in it and cover it over with ice.

The best time to harvest onions is when the tops turn yellow and break over just above the bulb.

On a warm sunny day, pull the bulbs, loosening them with a spade if the soil is too firm to release them easily. Allow them to dry several hours in the sun. Then cut off the tops about an inch above the bulb and remove any dirt. Put the onions in small open mesh bags in a well-ventilated dry room, porch, or garage. The firm storage type onions will keep for several months in a dry location.

Leeks are very useful when a mild onion flavor is desired. They require a long season of more than four months to mature. In late summer, mound up dirt around your leek plants to a depth of six to eight inches. This will blanch the basal portion and improve the quality. Harvest beets when small to medium size. Large ones are likely to be tough and woody.

Beets grow best in cool weather, and they can be harvested a relatively short time after planting. The greens are better when picked small; they get tough as they grow older.

Pick snap or string beans when the pods snap readily.

Pick broccoli before the flowers show color, although an occasional opened flower does not mean overmaturity.

Cabbage heads should be firm and heavy for their size.

Carrots contain the most sugar when fully matured, but they are better eating when small, firm, and well colored.

Moderate-sized cucumbers are best. Yellowing indicates age.

The Summer Garden

Freezing or canning your winter and spring harvests will make the specter of a tropical summer less threatening if you are attempting to grow most of your own vegetables.

Tropical summers were never designed for the growing of temperate zone vegetables—although they are ideal for dozens of other delicious edibles. Bugs are in their glory. And the rainy season downpours, coupled with high humidity, don't spare less-hardy vegetables.

Some people have success with the so-called summer squash, planted in March. The small cherry tomatoes (there are several types like Burpee's 'Pixie,' 'Patio,' and 'Tiny Tim') may do well, especially if potted and watered well. Cantaloupes and watermelons are usually planted in March, along with okra, chard, pole beans, Southern (black-eyed) peas, mustard, and bulbing onions (onion sets) for summer harvest.

Tampala is a classic summer plant in South Florida, although not widely grown. Known as "summer spinach," it has a flavor different from other vegetables used as "greens." Its leaves are cooked, prepared, and served like spinach. Most seed companies carry it. New Zealand spinach is another offbeat but obtainable summer vegetable.

Eggplants can make it through the summer with plenty of food and water. Bell peppers should be given partial shade and sprayed, along with eggplants, with a combination insecticide-fungicide. Most summer vegetables do better with a little shade.

Cucumbers are grown with varying degrees of success during the hot months at the tip of the state. 'Poinsett' and 'Ashley' are resistant to downy mildew–which strengthens their chances for success–but they, too, need regular spraying.

The Tropical Vegetables

Chayote, pigeon peas, calabaza, malanga, dasheen, cassava, boniato, sweet potato—these are some of the delicious vegetables enjoyed in the Bahamas, the Caribbean, and in Central and South America but denied to South Floridians until the influx of Cuban immigrants made them more accessible.

Now the South Floridian with an adventurous palate can grow these much-easier-to-raise vegetables here. And most of them produce year-round, summer as well as winter. They are fascinating to learn about—each has its own set of unusual characteristics—and add a new look to the landscaping, as well.

For awhile after they became available here, only the Latins who knew and loved them had any information on how to care for and cook them. Now, however, with several helpful publications, the Dade County Cooperative Extension Service is prepared to help non-Spanish speaking homeowners grow and serve them.

These vegetables may be obtained from Cuban markets—a number of which can be easily located along Southwest Eighth Street in Miami—and planted either as seed, tubers, or "eyes." Sometimes, too, they are advertised in the "Market Bulletin" published by the Florida Department of Agriculture and Consumer Services, Mayo Building, 407 Calhoun Street, Tallahassee, Florida, 32304. Subscriptions to the bulletin are free and so are its classified ad pages where hundreds of gardeners find—and advertise—unusual plants and vegetables.

Because some people with timid taste buds are reluctant to try growing new foods, one of several ways to prepare each vegetable is presented, following its description and cultural hints. There are many more ways to use them, and dozens more for boniato alone are included in the free booklet, "Recipes for Dade Grown Boniato," available through the Dade County Cooperative Extension Service office in Miami and Homestead.

Boniato and sweet potatoes: Long a favorite vegetable in Latin America, the sweet potato *(Ipomoea batatas)* is grown throughout the West Indies as at least 88 distinct cultivars. The boniato is a very fine, white-fleshed sweet potato with fewer calories and a slightly different flavor than those with which we are most familiar. Over 800 acres of boniato are grown in Dade County, and the vegetable is available almost all winter long.

A frost-free growing period of four to six months is needed for sweet poatoes, along with full sun to light shade and moderate but consistent moisture. Good drainage is important. They often are planted (not mandatory) on foot-high or more ridges three to four feet apart with the cuttings a foot apart. They like manure, but if too much nitrogen is used large tubers don't develop.

In the tropics, sweet potatoes are propagated by stem cuttings, 12 to 18 inches long, taken from mature plants. The bottom leaves are removed and the lower half of the cutting is inserted in the soil at an angle. Sometimes the tubers are allowed to sprout on windowsills, after which the sprouted sections are planted.

Time to maturity is three to six months. The leaves will turn yellow and begin to drop. Or you can test a tuber. If, when you cut it, the sap dries up rapidly without discoloration, it's mature.

Weevils are about the only significant pest. If they do show (as small white grubs in the tubers) it is best to destroy all vines and not plant tubers again on the site for a couple of years.

A recipe for ORANGE CANDIED BONIATO:

1 cup orange juice
1 cup water
1 cup sugar
1/2 teaspoon salt
1/2 teaspoon grated orange rind
6 medium sized boniatos
1/4 cup butter

Peel uncooked boniato, cut into 1/4-inch slices, and arrange in a baking dish. Make a syrup of the remaining ingredients and pour over potatoes. Cover and bake in moderate oven until tender.

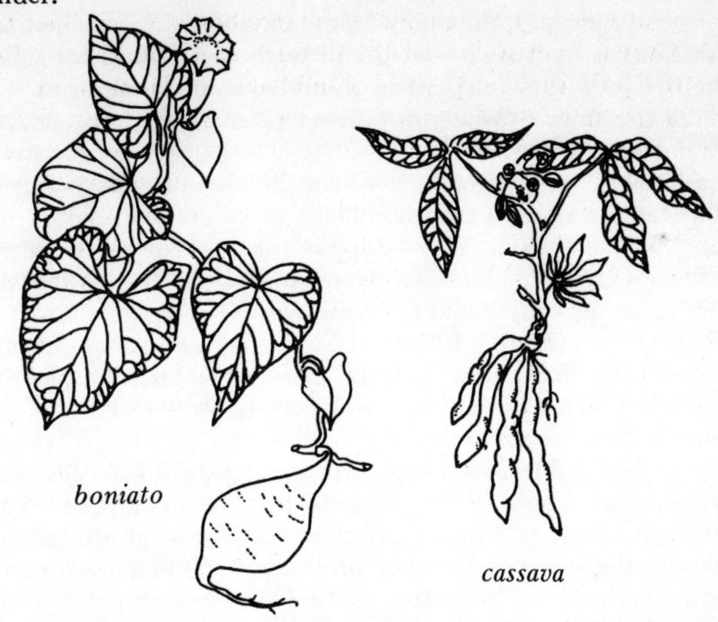

boniato

cassava

Cassava (also called yucca, guacamote, manihot, manioc, tapioca): Native to western and southern Mexico, parts of Guatemala, and northeastern Brazil, *Manihot esculenta* has a colorful history as a staple in the diets of tropical residents from ancient to modern times.

Starch is made from it by grating or grinding washed, peeled tubers and washing out the starch by squeezing in repeated changes of water. Tapioca is made by gently heating washed and cleaned starch on hot iron plates; this partly cooks it and makes it form small round pellets. Cassava flour is made by grinding the sun-dried slices or chips. In northeastern South America, a coarse meal, farinha, is prepared by grating washed, peeled tubers and squeezing the mass in long, sleevelike baskets to release the juices. The compressed pump is then toasted over a low fire. The latex and extracted juice may be concentrated by boiling to produce "cassureep," used in sauces and as an ingredient of the famed West Indian pepper pot. The leaves are used as a potherb, especially in Africa.

A shrubby perennial from six to seven feet tall, it comes in bitter and sweet types. The sweet types are grown mainly for eating as vegetables; the bitter types contain a higher percentage of poisonous hydrocyanic acid and are raised mainly for flour. (The poison is dissipated when the tubers are grated for extraction of the flour or when they are pared and boiled or roasted.)

Relatively drought-resistant, cassava is a heavy feeder and a potash-lover, preferring a fertilizer in the neighborhood of 4-8-10 at the time of planting. Once or twice again during the year the same fertilizer may be applied as a side dressing. Tubers are harvested within nine to twelve months after planting, depending on the variety and the weather.

Propagation is by 10- to 12-inch mature stem cuttings planted three feet apart in rows four feet apart. Put them in almost a horizontal position with about three-fourths of the cutting below the soil.

Since all sizable root crops appreciate a deep soil, you may want to plant yours on a mound if you have shallow soil. Further, mounding soil for several inches up the stem after about two months encourages tuber production. Tubers are usually dug individually as required, as they begin to rot within 48 hours after being taken from the ground.

Pests are minor on cassavas, and the only serious diseases are viral, for which there is no control. Some scale insects and spider mites have been known to visit them.

A recipe for CASSAVA PUDDING:

1 level cup grated cassava	2 cups milk
1 3/4 cups sugar	1 egg
1 teaspoon salt	3 cups water
1 tablespoon butter	1/2 teaspoon nutmeg

Wash, peel, and grate freshly dug tubers. Mix with milk, sugar, egg, salt, water, and butter, and flavor with nutmeg. Bake 1 hour in buttered baking dish.

Chayote (also called choyote, christophine, chuchee, miriton, "vegetable pear"): A delicacy native to southern Mexico and Central America, favored by the Aztecs. The chayote itself is the gourdlike fruit of a trailing vine *(Sechium edule)*, a cucurbit like the temperate-zone squash, which it resembles in taste.

The fruit is eaten boiled as a vegetable, as are the large tuberous roots. The young leaves and tender shoots are sometimes eaten like spinach. Chayotes will bear fruit from fall to spring, but especially in the fall. It is a perennial—unless killed by frost. (If very cold weather threatens, bank the base with soil.)

The whole mature fruit may be planted horizontally and thinly covered with soil. Or it can be planted vertically, stem end exposed. Several can be spaced four to six feet apart along a fence or trellis. Harvesting of fruit may start three to five months after planting. Fruit should be picked when about two or three pounds in size, still immature and before the seed is well formed. The pale green flesh when cooked has almost the texture of butter and the taste of a delicate squash.

Chayotes should be planted in full sun in nematode-treated soil. As with other curcurbits, powdery mildew can be a problem unless the plant is sprayed weekly with a fungicide. A well-drained, well-fed soil increases production, especially with a heavy mulch.

A recipe for BOILED CHAYOTE:

3 chayotes	black pepper
1/2 teaspoon salt	butter

Wash, scrub, cut in half, and peel chayotes. Cut in quarters of smaller pieces. Place in pan filled with one inch of salted water and bring to boil, turn flame to low, and cook gently for about fifteen minutes. Test with fork for tenderness. Drain, add butter and pepper.

Dasheen: The names—both common and botanical—of edible tubers belonging to one of the elephant-ear genera, Colocasia, are constantly disputed, even by experts. The dasheen and taro are both regarded by the authority Purseglove as *C. esculenta* var. *esulenta,* and the eddoe as *C. esculenta* var. *antiquorum.*

West Indian dasheen cultivars have a large central corm, which is the main edible portion, and few side tubers or cormels. West Indian eddoe cultivars have a relatively small main corm and many side tubers, which constitute the main edible portion. The dasheen of the southern United States is the eddoe of Trinidad.

Dasheen are grown from small side tubers or suckers in a wet, heavy soil. A fertilizer high in potash can be applied in two or three applications over the eight months or so it takes tubers to mature. Leaves will begin to yellow and die down at maturity, when the large central corm may be pulled and eaten, the side tubers planted.

When planting dasheen it is a good idea to avoid nematode-infested soil. Blights, viruses, and leafhoppers have been reported at various points around the world on this plant, but for the backyard grower its culture is easy. White grubs and wireworms generally do minimal damage. In dry weather, aphids and red spider may appear on foliage.

Properly cooked, dasheens are mealy and have a nutty flavor. In food properties they resemble the potato, but they are richer in both protein and starch. A four-ounce serving has slightly more than 100 calories.

Recipe for BOILED DASHEEN:

Wash thoroughly with a brush and cook without peeling. Raw juice is irritating to the skin. Cover dasheens with boiling salted water, and cook from 15 to 30 minutes, or until tender. Dasheens require less cooking time than potatoes. Peel, cut in cubes, and season. Serve in pieces or put through a ricer.

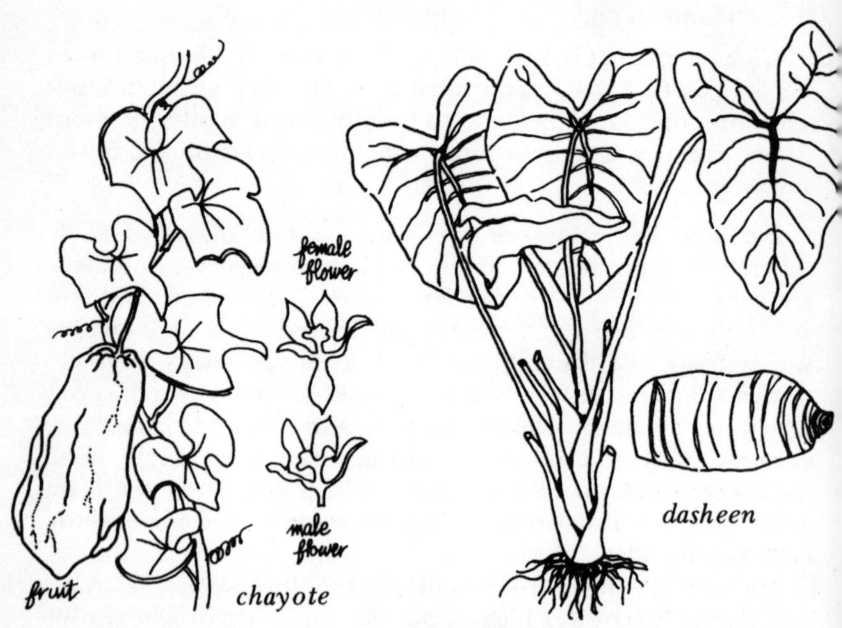

Eddoe (See dasheen description): These plants have tubers that resemble the dasheen in taste and may be boiled, roasted, or baked. The young leaves can be used as spinach, but they are more acrid than the dasheen's. Sometimes known as the "Chinese eddoe," it was brought to the Caribbean by Chinese immigrants. In Spanish-speaking parts of the Caribbean they are sometimes called "malangas," although this term more often refers to other aroids, including xanthosomas. Eddoes are hardier than dasheens and can be grown with less rainfall and on poorer soil. They can also withstand colder weather. They are usually planted in large holes, and organic manure may be added. They are replanted every year after a five to six months' maturity. Unlike dasheens, eddoes can be stored for several months. Problems could be the same as with the dasheen.

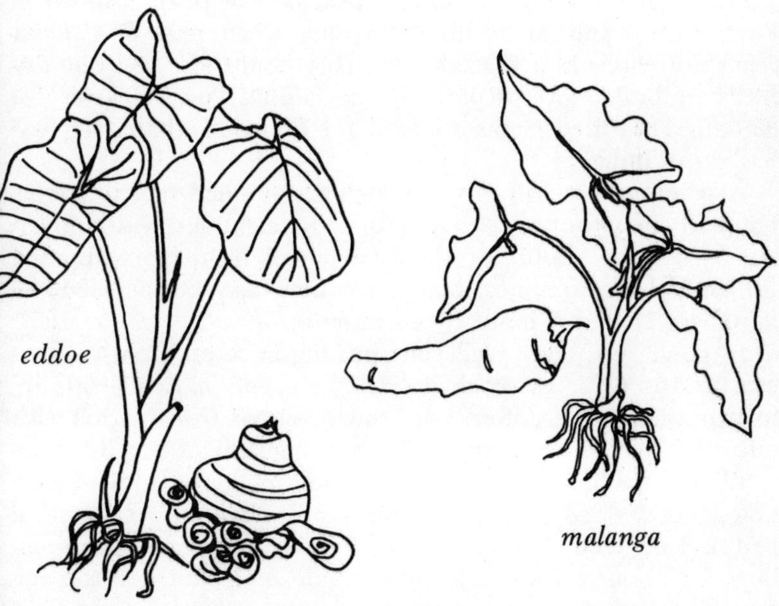

Malanga (tannia, tiguisque, yautia, cocoyam): The malanga as it is known in Cuba is *Xanthosoma sagittaefolium*. In South Florida, the type grown is a "blanca," probably from the Dominican Republic. (Malangas originated in Africa.)

Highly prized by Latins, it produces cormels around the

mother plant, ranging in size up to more than half a pound with a length of about five inches and diameter of an inch and a half at the terminal end. Eight or ten cormels may be produced per plant in about eight months. If protected from drying out, the cormels keep well at room temperature.

The plants can reach seven feet and should be planted at three-foot intervals in rows twelve feet apart. They like a 4-8-6 fertilizer, applied at a rate of a quarter pound per plant monthly. A nutritional spray of chelated iron applied to the ground may be advisable if plants begin to show green veining on yellow. They can be watered twice a week during dry periods and protected from frost with a sprinkler.

Watch for same pests as on dasheens.

Malangas can be prepared as dasheens are.

Pigeon pea (red gram, Congo pea, no-eye pea): Canned in Puerto Rico and Trinidad, the young green peas of *Cajanus cajan* are eaten as a vegetable in many countries. The ripe dry seeds are boiled and prepared in the famous "peas and rice" of Nassau. The dried stalks are used for firewood, thatching, and baskets in India.

A perennial, it will grow as a shrub or small tree to twelve feet with a six-foot or more limbspread, height and width differing by variety. Color of the seed ranges from ivory through brown and red to almost black, and they may be speckled and blotched. There are hundreds of varieties.

In good soil plants will start bearing in seven months. Seed can be stored for two years to insure a supply in case plants are lost to a light freeze. Seeds are sown four or five to a hill with hills about eight feet apart. A small handful of a 6-6-6 type fertilizer should be mixed with the soil in the bottom of each hole, and covered with a little plain soil. Then the seed is dropped and covered with an inch of soil, well firmed. When seedlings are about a foot high, they should be thinned to the one that is strongest. Young plants are slow growers and make their most luxuriant growth after about four months. Small amounts of 4-8-6 will help the plant as it matures to form more seed than leaves. It is drought-resistant and does not need constant watering.

Although wind-resistant, pigeon peas should be planted,

ideally, on the southern side of buildings to protect them against northern winds in climates where there is the chance of frost.

The plant experiences few pests in Florida.

Recipe for PIGEON PEAS AND RICE:

1 cup dried pigeon peas soaked overnight and drained, or 2 cups freshly shelled green peas	1 cup tomato sauce
1 lb. lean pork, diced	1 bunch chopped parsley
½ cup salad oil	Bay leaves
1 chopped onion	2 tsps. salt
	½ tsp. black pepper
	¼ cup rice, washed and drained

Cover peas with fresh water, add diced pork, salt, pepper and chopped parsley; cook over a low heat until tender (about two hours for dried peas), adding hot water as often as necessary. Meanwhile heat the salad oil in a saucepan, add the chopped onion, and sauté for 10 minutes, stirring often.

Add tomato sauce and bay leaves. Simmer five minutes and pour the mixture into the pigeon peas. Add 1 1/2 cups hot water and bring the whole to a rolling boil. Stir in the rice and cook 25 minutes or until rice is done. Watch carefully and add small amounts of hot water to prevent burning if necessary.

Roselle (Jamaican sorrel, Florida cranberry): *Hibiscus sabdariffa* is an annual shrublike plant from four to seven feet or more tall, with stiff, upright reddish stems. The alternate green leaves are "entire" on young plants, palmate with three to five lobes on older growth. The flowers, which are about three inches wide, are yellow with crimson centers; they usually appear in October. The bright, huge, red calyces around its fruit are the edible parts. These mature usually in November, but the ripening season may be extended several months by harvesting all of the fruit promptly.

Propagation is by seed. Plants are sown in March in nematode-treated soil that is not too shallow. Plants can also be propagated by cuttings. Space four to six feet from other plants.

This plant benefits from manure, but after young plants are started, a 6-6-6 applied lightly a couple of times a year will provide all the nutrients needed. Overfertilizing is not desirable.

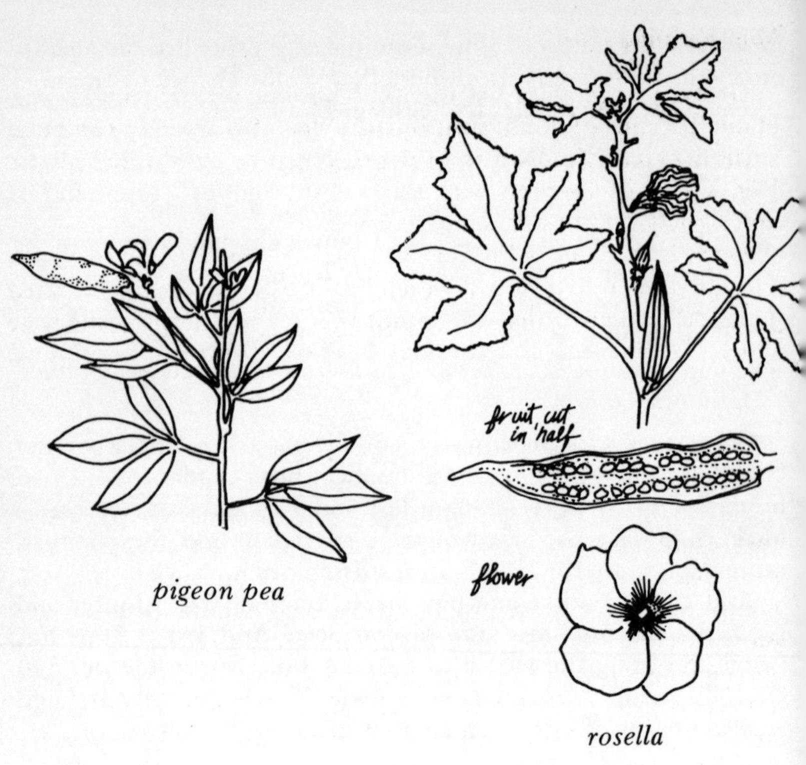

pigeon pea

rosella

Tender young leaves and stalks can be used in salads or as a potherb. The most serious pest is the caterpillar.

Recipe for ROSELLE SPONGE:

1 quart roselle calyces	3 tablespoons unflavored gelatin
1 pint water	1/2 cup cold water
1 cup sugar	

Stew the roselle in the pint of water until soft. Rub through a colander. Add one cup of sugar and return to the stove. Soak the gelatin in the half cup of cold water and add to the hot mixture. Pour half of this mixture into a wet jelly mold and chill until set. Chill the remainder until partly stiffened, then whip with an eggbeater until it is a mass of fluff. Pour this mixture lightly over the first part. Chill until firm and serve with orange sauce.

Mini-Farming

Growing vegetables has naturally become identified with vegetable patches of some size, but the "patch" actually can be as small as a patch on a pair of pants. It can be a soil-filled plastic bag, a ring of chicken wire filled with compost, a pot full of gravel, or a barrel punched with holes.

One form of mini-farming has become so popular in South Florida that nearly every other backyard contains a so-called Japanese Tomato Ring, a tidy unit that turns out 100 pounds or more of tomatoes in seven feet of space—for a total outlay of $5 to $10.

The Japanese Tomato Ring

Granted, all it is is a compost pile surrounded by tomatoes. Gardeners have been putting stray vegetables around compost piles ever since compost piles were invented. But this handy little unit is responsible for getting thousands of people more interested in the wonders of compost and the joy of producing their own food. The results are dramatic, and virtually everyone succeeds the first time. Basically, this popular project involves growing four tomato plants around a circle of wire fencing. The tomatoes are planted in ordinary garden soil outside the ring but their roots head under the big wire "feeder" full of goodies that are continuously released.

Ingredients for the ring include: a length of wire fence five feet high and about 15 feet long; a small bag of all-purpose garden fertilizer; about two wheelbarrow loads of good soil; a small bottle of nematode killer, available from any garden supply store; a quantity of mulch; four tomato plants. Three plants might produce as much fruit, but four help cover the wire frame more quickly.

Pick a location in full sun and protected from north and northwest winds, if possible. Clear a circle seven feet across and break up the soil to a depth of several inches. Treat this area with the nematode killer, following directions on the label.

Arrange the fence in a circle, which will be about five feet in diameter, and place it in the center of your cleared ground.

Put a layer of mulch about six inches deep in the ring. Add a

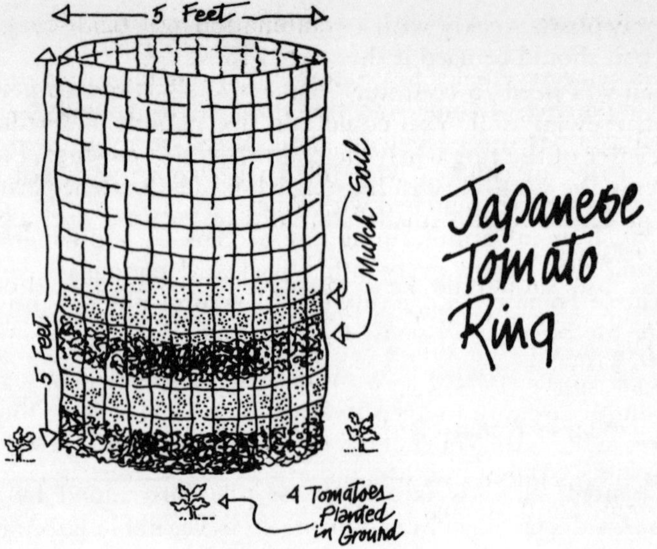

layer of good soil, another layer of mulch, and a final layer of soil. If you have no mulch, you could use a three-inch layer of peat moss.

Shape the top layer of soil to form a shallow dish. This will receive water after the plants are well started.

Add about two pounds of fertilizer to the soil and mulch. Then treat the mulch and soil with more of the nematode killer to help prevent root-knot.

Wait a week after preparing the ring before you plant. Then set no more than four plants in the cleared soil outside the bottom of the ring. Fertilize the soil around the plants, but do it lightly to avoid burning the plant roots.

Water the area around the ring when the plants are small. Apply water to the dish-shaped pile of soil and mulch in the ring after the plants get larger.

The roots of the plants will be attracted to the wealth of plant food inside the ring. Tie the plants to the wire with soft cloth or other material that will not damage the stems. It won't be long before the plants cover the wire and fill the inside of the ring. You may have to prop up excess growth.

Spray plants weekly with a combination insecticide-fungicide. Slug bait should be used if these pests show up.

You will need to continue fertilizer applications to keep the plants growing well. You could add five pounds of fertilizer to the center of the ring when the plants begin to produce, but this presents the possibility of burning roots. The gardener probably will do better to add smaller amounts of fertilizer over a period of several weeks.

The soil should be kept moist, but not wet. A thorough watering once a week in dry weather should be enough, but it will depend on your soil, wind, and other conditions.

Some gardeners lace a two-foot strip of screening around the bottom of the ring to keep any mulch and soil from falling out. Others add extra touches like covering the fence with green paint. This prolongs its life and makes it less conspicuous until it is draped with its superproductive vines.

Strawberries in a Barrel

The traditional barrel for this old-fashioned form of strawberry raising was a large, strong, well-wired wooden barrel. If such an item can still be found, it makes a very attractive device, although the insides should be painted with a polymer or epoxy-resin to prevent rotting.

There are alternatives to the old wood barrel. One is a brown imitation wood-grained barrel made from plastic. As "unnatural" as it may sound, it is very practical with holes cut in the sides and bottom—and very attractive when planted. You can even use a brown plastic trash container, since most of it will be covered anyway. And even a small mesh wire "cage" similar to the Japanese Tomato Ring is feasible.

Bore or cut two-inch holes in whatever you use, in rows six inches from the top and covering the container, six inches from one another and six inches from the bottom. Stagger the rows so that there is a space in the row above each plant.

Drill a half dozen small holes in the bottom for drainage. Remove the head from the barrel (if you haven't already) and situate it in an open, sunny spot. Set the barrel on a low, slatted platform or bricks for best ventilation. (Some people like to put their unit on rollers so that it can be relocated for a different summer crop.) Next, place a two-inch perforated PVC (plastic) pipe in the center for watering-feeding, holding it in place as you put in the soil.

Your mix should be a sterilized potting soil available in bags from any garden shop. Usually, this mix will contain soil plus peat moss and perhaps perlite with a manure added. If it doesn't have peat moss added, you should incorporate peat moss to equal about one-third of the volume of mix. A chemical fertilizer should not be added as it will burn the plant roots on direct contact.

Add the soil to just below the first line of holes and water it well. Then, with great care, put the plants inside the barrel and bring the foliage out through the holes. If the barrel or cage is too tall to reach down inside, you'll have to work the roots carefully through the holes.

Pour in more soil until the level reaches just below the next row of holes, then water down the soil so the roots are well settled. You may have to add more soil before you repeat the process with the next row of plants.

To help keep the dirt from escaping from the holes, you can stuff some sphagnum moss into the openings or use discarded women's hosiery spread over the soil and caught inside the hole. Coconut palm fiber also works well, but keep whatever material you use from touching the stalk of the plants. Place a few extra plants on top of the barrel for replanting if needed.

Some people use liquid manure to get the plants started, but if there is manure in the planting mix, you won't want more. A 20-20-20 water-soluble fertilizer poured into the pipe every other week will give the plants the nutrients they need. If you can find a liquid fertilizer higher in phosphate and potash (the second and third numbers in the analysis) than nitrogen, your plants will do better.

Use an insecticide-fungicide spray weekly, and your unit will be an admirable sight from October through April. Most people plant strawberries in October-November and again in February.

Barrel culture for strawberries is not only "cleaner" but lets the grower pick standing up, a definite advantage.

Container Farming on Balconies and Patios

You may not be able to feed a family entirely on what you can produce on an apartment balcony or townhouse patio, but you will have the fun of watching "useful" plants grow, plus the special flavor and interest of homegrown vegetables to help fill out your meals. Container gardening is also a fine way to introduce youngsters to the glory of growing their own food.

You'll need an area that has direct sunlight (southern exposure is best, by far), enough support for heavy containers, easy access to water, and protection from strong winds.

Anything that holds dirt and doesn't dissolve can be used as a planter if you add drainage holes. (Three quarter-inch holes per square foot of bottom is a good rule of thumb.)

Take wood bushel baskets discarded by markets, line them with a plastic garbage bag and pour in potting soil, punching holes in the plastic. Buy plastic trash cans, drill drainage holes, and wrap them with reed fencing cut to size. Those brown imitation wood barrels made of plastic and used for large waste baskets are also attractive.

The Sunset Magazine's series of garden books, available at almost every department store book section and garden shop, includes a big $1.95 paperback called *Gardening in Containers,* which has many well-illustrated ideas for handsome, expensive-looking, do-it-yourself containers. (Take it to a high school woodshop instructor and ask him if he or a student would make it for you, if you're not handy.)

If your container is plastic, drill along the sides near the base to keep it from weakening. To keep holes unclogged, line the planter bottom with an inch of gravel. A sterilized soil is more expensive than untreated topsoil sold in unmarked plastic bags, but it eliminates a lot of fungus and nematode problems that might be present.

You could take a chance with untreated soil or treat it with Nemagon, available at any garden shop. Untreated topsoil usually works well. Ideal is a mixture consisting of a third black dirt, a third coarse builder's sand or perlite, one-sixth peat moss, and one-sixth processed or dried manure.

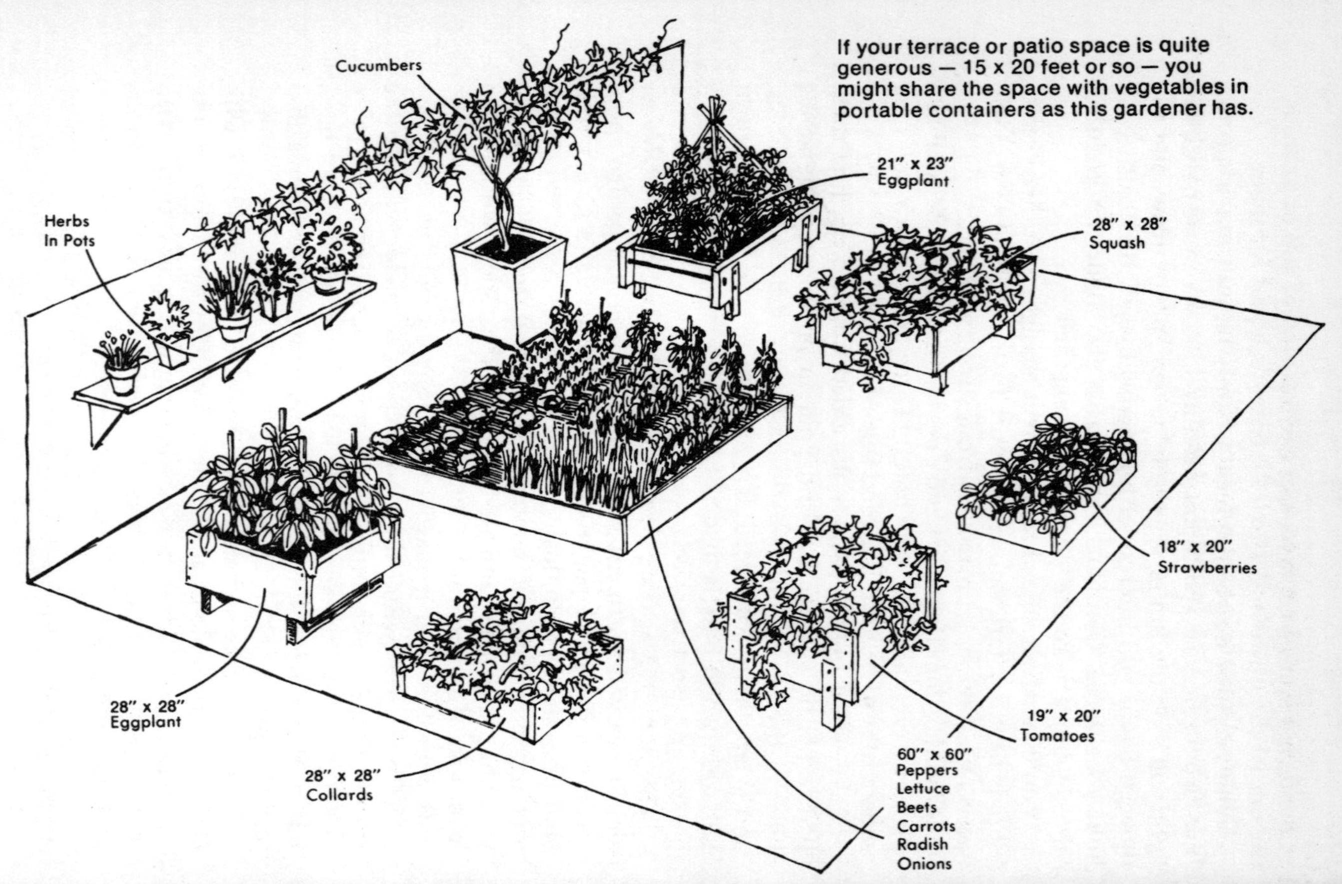

If your terrace or patio space is quite generous — 15 x 20 feet or so — you might share the space with vegetables in portable containers as this gardener has.

A dry 6-6-6 can be mixed into the top inch of soil or you can use one of the powdered 20-20-20 types as liquid fertilizer.

Container plants should be watered when you can stick your finger about an inch into the soil and still hit dry or just slightly moist dirt. Be sure the container drains well. Four hours of direct sunlight daily (six for fruit-bearing plants) is the absolute minimum most plants need to grow adequately, and six to eight is better and will give you more to eat.

To control pests with a dust insecticide, use it weekly in the early morning when dew will help hold it on the leaves. You might also want to combine your vegetable plantings with natural pest-controlling plants like marigolds and nasturtiums, which also will add color and fragrance.

If you work and there is no one home to water on very hot days, you might want to consider setting your containers on blocks or bricks, putting a tray of water underneath and running a wick from the tray into the holes.

Containers can be mulched with sawdust, vermiculite, rice hulls, oak leaves, or any other fine material that will help retain moisture.

Tomatoes are a very popular container plant because they produce the highest yield. The best plant to buy is stocky and bushy—and is one of the varieties that does well in South Florida. Prune it down to one stem, set it in the soil, and tie it loosely to a stake. Pinch off suckers (useless growth where branches join the stem) at the tip after they've formed at least two leaves.

For full-sized varieties, use a five-foot tall circular cage of six-inch wire mesh for support. If you let one of the suckers develop into a second stem, you'll get more tomatoes and the extra foliage will provide better sun protection.

To help cut blossom loss due to unfavorable temperatures, use a hormone spray called "Blossom Set," available at your garden shop. And if cool, damp conditions interfere with the setting of fruit, you may need to help the plant release its pollen by whacking at the stake or shaking the cage during the warmest, driest part of the day.

Other plants that look good in containers are herbs, eggplants, peppers, lettuce, beets, carrots, radishes, onions, strawberries, cucumbers, squash, and spinach.

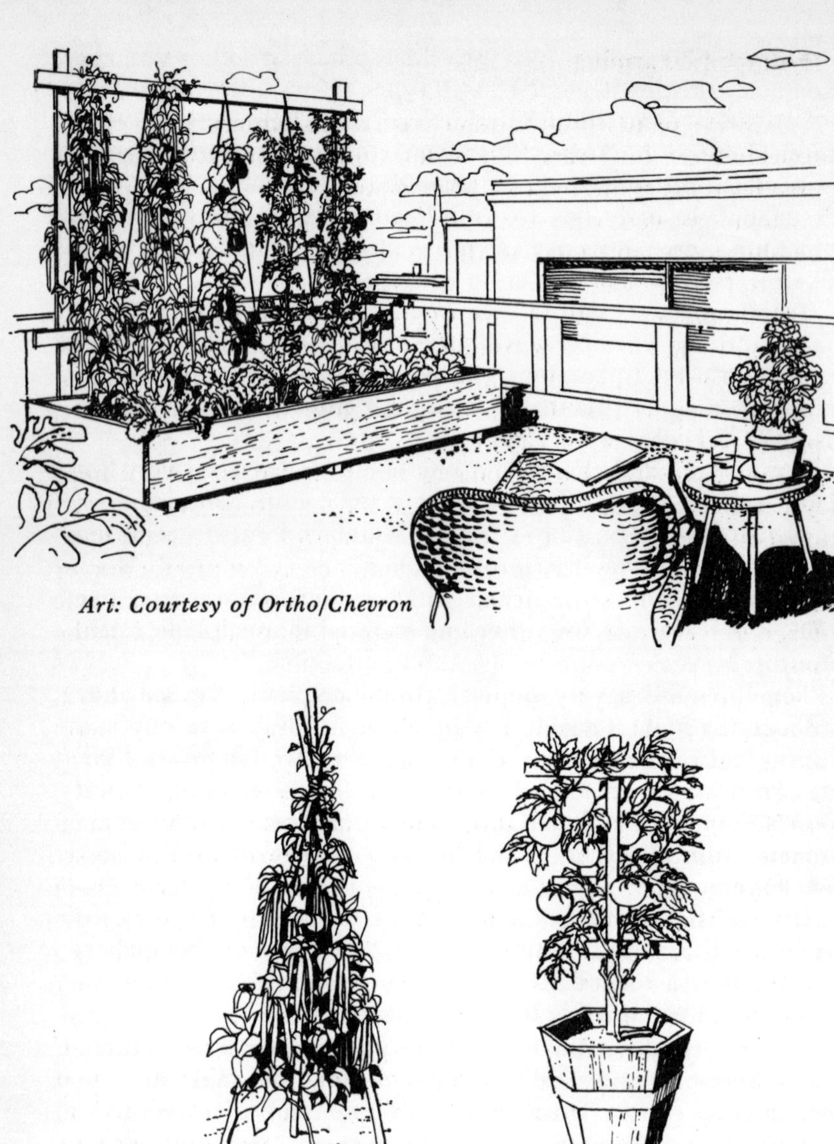

Art: Courtesy of Ortho/Chevron

Fruit-bearing vegetables can be kept pruned to supports and grown in containers on rooftops, balconies and patios. Beans, eggplants, cucumbers, and tomatoes will thrive with extra attention to watering and some wind protection.

Hydroponic Farming

Growing plants without soil has fascinated large- and small-scale farmers for years but, at least on a commercial basis, its practicality is restricted to a few isolated cases.

Home growers who try it should be homebodies—or should be able to rig up timers so that two or three daily trips to the unit to feed the plants are not necessary.

Hydroponics is the science of growing plants with their roots suspended in a solution containing plant nutrients. It could be called gravel culture, also, since this is the most popular medium used to support the plants. Also used is sand, cinders, sawdust, perlite, vermiculite, or well-rotted plant material.

The claims for hydroponic growing are legendary—turnips growing to baseball size in 45 days, flawless tomatoes up to 25 ounces each, a head of cabbage weighing over 10 pounds, dahlias 44 inches in circumference, and an increase in production of four or five times the ordinary. These statistics are no miracle when you consider the attention hydroponics calls for, but the results make everyone happy who has the time.

As a container you can use a flower pot, a tin can, a nail keg, a half-barrel, or a bushel basket. Even styrofoam ice chests are being used successfully. If you write to Dr. Chatelier's Plant Food Company, P.O. Box 20375, St. Petersburg, Florida, 33742, they'll send you a free sheet illustrating a simple, inexpensive "sub-irrigated" hydroponic garden box you can make.

All containers used for hydroponic growing must have a hole cut into them about an inch from the bottom on the side of the container. A plastic drainpipe should be inserted and puttied to seal the hole. If you are using any supportive medium other than shavings, it's a good idea to cut a two-inch piece of window screen to place over the drainage hole inside the container. This keeps the medium from being washed out every time you feed your plants. (Bushel baskets have built-in drainage and won't need this touch.) A small bucket can catch the overflow.

As a medium, vermiculite and perlite can be used together in a 50-50 ratio or separately. The coarser the quality, the better. Before planting, this material must be soaked well.

Untreated pine or cypress wood shavings come highly recommended. Soak them down with a teaspoonful of plant food per

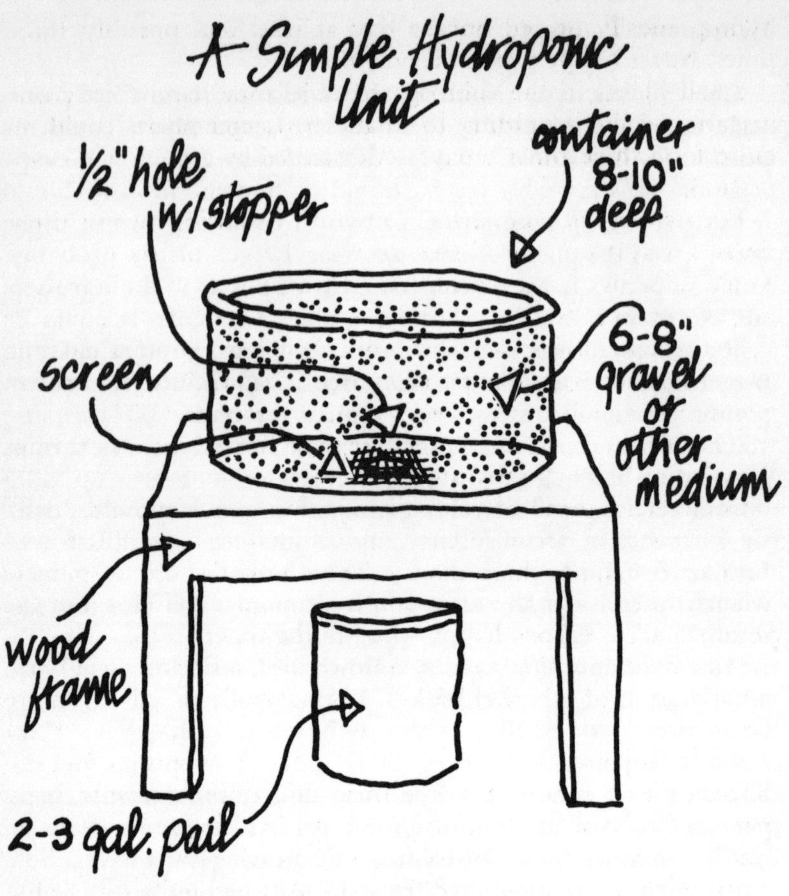

gallon of water for a few days before planting. This helps the chips absorb additional plant food quickly. Be sure drainage is good. Wood shavings are not suitable for the subirrigation method of hydroponic growing.

Regardless of the growing medium used, gravel in the bottom of the container is the old standby to promote good drainage. It can also be used as the only medium.

How many plants can you pack into a unit? Tomatoes need a square foot each to produce best. Other plants need less. The depth of the container doesn't need to be over eight inches, as long as the square foot of space is there.

Now for feeding. Chatelier recommends that plants grown

hydroponically be fed once a day at least and possibly three times. Never let the plants become dry.

Small plants in the subirrigated units may require only one irrigation daily, according to Chatelier. Larger plants could require food three times a day, as demanded by growth and evaporation.

For surface-fed containers, a cupful of solution two or three times a day promotes faster growth. Larger plants probably would need two or three cupfuls, depending upon their growth rate and evaporation.

Mixing the plant nutrients and keeping the elements in proper balance is important in hydroponics. Unless you are experienced with chemicals, don't try to formulate your own nutrients. It's much easier to purchase ready-mixed ingredients for the solution at any garden supply store.

Commercial fertilizers may contain insoluble materials. Possibly a quarter of the fertilizer may not dissolve. Don't use any fertilizer containing more than one percent chlorine.

For planting stock, use only the recommended varieties for South Florida. Rinse off the soil or other materials that cling to the roots and spread them out on the gravel, covering them with another layer of gravel as high as soil normally would reach up the stem.

Sturdy supports are necessary for most of the plants that do especially well in hydroponic culture—tomatoes, eggplants, peppers, and cucumbers. Spraying for bugs and diseases is done as usual.

Many garden shops carry the helpful booklet on hydroponics distributed by Nutri-Sol, a water-soluble fertilizer manufacturer. The public library is also a source of hydroponic information.

Herb Gardens

There is a mother lode of material nowadays on herbs since they place such an important role in the current revival of interest in exotic recipes. But there is very little solid information on growing them, particularly in the tropics where they cannot always be raised as annuals.

Herbs don't like hot weather and humidity. Most prefer the

  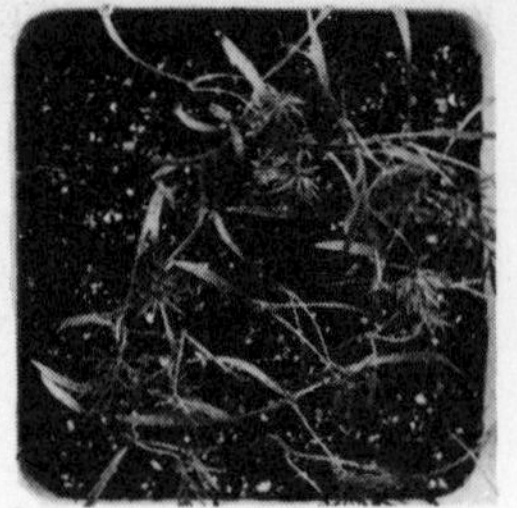

Photo: Michael Lloyd Carlebach

Herbs are grown as annuals in South Florida since they rarely survive the hot, rainy summers. They sometimes can be purchased locally as young plants but most frequently are grown from seed ordered through seed company catalogs.

coolest months and are grown as annuals in pots with a sterilized soil, peat, and perlite mix. Here is a review of the cultural conditions various herbs favor. (Include them in your spray program for vegetables only if you take care to observe days-to-harvest warnings on the pesticides-insecticides):

Anise. Growing season for this 18 to 24-inch beauty is about four months. When seedlings reach two inches in height, remove smaller ones so that plants stand about eight inches apart. Soil piled around weak stems will help support them. When the seed heads turn pale gray, harvest by cutting them off.

Sweet Basil. It grows 15 to 24 inches, and a single mature plant will produce three cups of fresh leaves in a season. It needs full sun and compost or manure. Set seed about a half inch deep. Seedlings should be spaced a foot apart, thinned to that distance when they are three inches high. You can start picking leaves when the plant is four to six inches high. For drying, harvest just as first flowers appear by cutting off stems four inches from the ground. To produce a second or even third crop after cutting, apply a 5-10-5 fertilizer at the rate of three ounces per 10 feet of row.

Borage. An 18-inch-high plant that can produce about 10 flowers and two cups of leaves for the first harvest. Leaves may be harvested again in one month. It needs full sun to light shade, doesn't like too much soil moisture. When seedlings are about two inches high, thin to a foot apart.

Caraway. Plants make a mound of feathery dark green leaves about eight inches tall the first year, and if you can get them to survive a second, they will send up flower stalks about two feet tall the following spring, dying after the seeds ripen. A marginal herb for South Florida.

Chive. These are good pot plants, doing well in a medium made from peat moss and perlite enriched with manure. From seed they take a year to produce. They do best in winter months. If plants weaken, scratch a light dusting of 5-10-5 into the soil. Chives multiply so rapidly that roots must be divided and reset regularly. Like onion sets they can be dug, stored, refrigerated, and planted the following season.

Coriander. In Latin America, this plant with the lemon-flavored seeds is known as "cilantro." It grows 12 to 30 inches tall. A single coriander will produce a quarter-cup of seeds over a two-month period or, if the flowers have been removed to prevent seeding, a half cup of leaves in the same period. It needs full sun, out of the wind. Leaves may be picked any time.

Because the seeds fall easily off the plant, it is necessary to drop the whole plant into a bag when the seeds begin to turn brown.

Dill. Cultivated for its seeds and leaves, this plant grows about three feet tall. A single plant produces a quarter cup each of seeds and leaves. It is harvested only once in South Florida. It needs full sun and does best in a neutral pH soil with compost or manure, away from the wind. Set seeds a quarter inch deep. If growth seems weak when plants are about a foot tall, feed lightly with 5-10-5. Pick the leaves any time. Harvest seeds same as with coriander.

Fennel. Fennel is one herb that doesn't mind an alkaline soil. Successive sowings at 10-day intervals are made by people who use it regularly. When plants are about a foot tall, feed lightly with 5-10-5. Leaves may be picked any time. Harvest seed when they begin turning brown.

Sweet marjoram. About a foot tall, a single sweet marjoram plant will produce a half-cup of leaves on first harvest, a quarter-cup at second cutting. It needs full sun, a light soil with alkaline pH. Leaves and stem tips may be picked any time for use fresh. When plants blossom, harvest the leaves for drying by cutting the stems off above the second set of leaves. New stems will come up after the first are cut. A good windowsill pot plant.

Mint. Mints aren't easy to grow in warm-weather areas. They are prone to fungus problems and should not be manured. If you can buy plants, you'll have more success than from seed.

Parsley. Parsley, which does well in South Florida, grows eight to twelve inches tall. It likes light shade and an alkaline soil enriched with manure. Soak seeds in tepid water for a day before planting to speed germination. Plant seeds a quarter-inch deep, and when they reach two inches, thin to three inches apart. When they grow to touch, thin again—every other plant. Repeat when they touch again, with the final spacing at 12 inches. When the plants are four inches tall—and again a month later—feed with 5-10-5. Leaves may be picked anytime.

A good windowsill plant.

Sage. *Salvia officinalis* reaches about two feet tall. Its small lilac-blue flowers attract bees. A single mature plant will produce a cup of fresh leaves. Sage likes a near-neutral pH. It can be grown from seeds but most often is purchased as a seedling. Leaves may be picked at any time for use fresh.

Shallot. A bulb of the onion family, shallots are grown for their mild garlic-flavored roots, which are made up of many segments called cloves, each with its own parchmentlike cover. Set out cloves about two inches deep and two to three inches apart. When the plants are five to six inches tall, feed with a 5-10-5 fertilizer. A month before the hot weather when they should be dug, fertilize with a high potash food.

Tarragon. A two-foot-tall licorice-flavored plant that produces about two cups of fresh leaves at first cutting, about a half-cup a month later. If you can buy small plants, do. After the first crop is picked, feed lightly with 5-10-5. Leaves may be picked anytime for use fresh. Florida is not the best place for tarragon but some gardeners do have luck with it.

Thyme. Thyme and lemon thyme can reach a foot in height. A single plant will produce a half cup of fresh leaves at each of two harvestings in a season. Since they take two years to reach usable size, these are best bought as plants. They are easily propagated from cuttings. Do not overfertilize. Leaves may be picked any time for use fresh; for drying, harvest leaves just before flowers open by cutting off half of the current season's growth.

Flowers to Eat

There are many common garden flowers that are not only edible but add spice and beauty to dishes. Before biting into flower arrangements, there are a few rules to follow.

Don't eat commercially grown flowers or flowers that have been treated with chemicals that are not approved for use on edible crops and therefore may have pesticide residue. Don't eat plants or plant parts that you do not know are edible. (*Sturtevant's Notes on Edible Plants* is a good guide to flower munching.) One plant part may be edible while another may not, so be sure the part of the plant you intend to use is mentioned.

Some of the most common edible flowers grown in South Florida include pot marigold (*Calendula officinales*), nasturtium *Tropaeolum majus*), roses, yucca, squash blossoms, and daylilies.

Chrysanthemums, not widely cultivated at the tip of the state because of humidity and heat problems, are a favorite flower of the Orient. In Japan and China both the flower and the leaves are eaten. The varieties grown here, however, have leaves too tough and bitter to eat. The flowers generally are used as an added touch in soups. Soaking the petals in iced lemon water will add to the flavor. Sprinkle the lemon-soaked petals over New England clam or fish chowder the next time you serve it for a unique and beautiful dish.

The calendula or pot marigold has been grown in herb gardens for centuries and is raised here as a winter annual. (It should not be confused with the common marigold.) The dried or fresh petals are used to color rice, noodles, soups, broths, cakes, salads, puddings, meat, cheese, and butter. Add a couple of dried petals to your next batch of deviled eggs, then form a flower out of fresh petals tucked into the yolks.

Nasturtiums have a flavor similar to, but not as strong as, watercress. The leaves and flowers are delicious in salads. The flower buds and green seed make fine capers. Serve nasturtium blossoms stuffed with your favorite spread for special occasions.

Roses have long been used in candies, cakes, spreads, and

jams. The old-fashioned and wild varieties are best in cooking. The petals make a tasty salad when mixed with chicory and served with oil and rose vinegar. Or serve a rosé wine and soda cooler along with rose petal ice cream and cupcakes.

Yucca blooms are sold in Guatemala as "Flor de Izote." Boiled or fried, they slightly resemble asparagus in flavor. They can be added to an omelet and served with a salad made from the heart or center of the bloom.

Daylily and squash blossoms may be used in about the same ways, although they have distinctively different tastes. The buds may be picked before they are fully open and boiled, fried, pickled, or used in omelets and fritters. Quick-fry partially open squash blooms and watch them continue to open as they cook. Daylily tempura goes well with chrysanthemum chowder. Try fried daylilies on your birthday; the queen of Thailand has them on hers.

Edible Wild Plants

Many plants native or naturalized to South Florida make an interesting eating experience, among them cocoplum *Chrysobalanus icaco)* which is canned in Cuba and the West Indies, sea grape *(Coccoloba uvifera)*, strangler fig *(Ficus aurea)*, *Hamelia patens*, prickly pear *(Opuntia austrina)*, and Seminole pumpkin *(Cucurbita moschata)*.

Best reference for wild foraging is *Wild Plants for Survival in South Florida* by Dr. Julia F. Morton, curator of the University of Miami's Morton Collectanea. (A number of the edible plants and trees are common in home landscaping.)

The Final Touch

No tropical or subtropical homestead is complete after the establishment of a flourishing vegetable patch unless the property is lushly planted with tropical fruit trees. Located well back from the plot so there's no danger of their casting shade across the sunny spot, fruit trees—carefully chosen—will help fill that summertime hole in the diet left by the exit of lettuce and other hot weather haters.

One of the wonders of South Florida is that it can be, most

A stunning ornamental tree that can be kept pruned to almost any height, the sea grape is famous for its dangling fruit, which is excellent for jelly. It can be grown in almost any soil, takes no care. It is native to Florida and the Caribbean countries.

days of the year, just as tropical as Thailand, Nigeria, Nicaragua, or Malaysia. Proof of the pudding is the colossal variety of exotic fruit trees that thrive here, from the macadamia and cashew to Mysore raspberries, jujube, and prickly pear.

There are many fruits (akee and canistel, to name two) that do not even taste like fruit, and these may be found especially nice in varying vegetable meals served to people who usually do not crave fruit. Beyond the wide variety of citrus, avocados, and mangos are dozens of tasty exotics, many of which grow on small-stature trees so you can plant more of them. (While on the subject of small, it should be noted that most people let their fruit trees get far too tall to care for or pick properly, and that a tree kept well pruned under 15 feet is much more inviting to pick.)

The yellow cattley guava makes an attractive small tree with tasty fruit that appears continuously but mainly in August and September. It is grown best in areas where its main enemy, the Caribbean fruit fly, is scarce.

If you contact your County Cooperative Extension Service for a copy of "Fruit Plants for Southern Florida," you'll have in your hands tabulated information on 107 exotic fruits. From this you can select fruiting shrubs with which you can build a "useful" hedge around your lot that will provide vitamins, minerals, taste, and interest yearlong.

Here are 10 fruiting shrubs/small trees that might catch your fancy:

Bignay *(Antidesma bunius):* A fine jelly, juice, and wine fruit from Malaysia.

Ceylon gooseberry (hybrid *Dovyalis hebecarpa* x *D. abyssinica*). Medium, yellowish-brown, sub-acid fruit, flavorful eaten fresh, in jelly, or as juice.

Grumichama *(Eugenia dombeyi).* A Brazilian shrub with small, purple-black, sweet fruit eaten fresh March through August.

Governor's plum *(Flacourtia indica):* From Madagascar, a sweet-fruited shrub with a round, smooth, red-to-black berry. Eaten fresh during the summer months. Plant two for company, pollination.

Barbados cherry, acerola *(Malpighia glabra).* Beautiful shrub covered with large, red, juicy berries very high in Vitamin C. 'Florida Sweet' is the best variety.

Strawberry tree, Cuban strawberry, Jamaica cherry *(Muntingia calabura).* Native to tropical America, this small tree bears delicious, sweet, aromatic red or yellow berries—good fresh, in jams or tarts.

Cattley guava, strawberry guava *(Psidium cattleianum).* Fine hedge plant with small, round, red or yellow subacid fruit from July to October.

Miracle fruit *(Synsepalum dulcificum):* From West Africa. Fruit makes sour things taste sweet. Produces irregularly all year.

Pineapple *(Ananas comosus).* Needs holes filled with good

Pineapples fruit well in Dade County where once, until labor costs soared, there were a number of commercial plantings. They do best in a soil enriched with humus and watered regularly. Full sun or light shade in summer is best.

soil mix, but the fruit needs no introduction. Dozens of delicious varieties available.

Sugar apple, sweetsop *(Annona squamosa).* Very decorative heart-shaped, segmented fruit with white, creamy, sweet pulp. Good fresh and in ice cream. A native of the tropical Americas.

Every genuine tropical homesteader also has a banana patch full of unusual varieties—or just the standards that taste so incredibly different from market types when fresh-picked. Where to find these fascinating fruits? That is the purpose of the Rare Fruit Council International, Inc., an organization of several hundred "rare" fruit growers in South Florida. If you miss or can't make it to their annual spring sale held in Miami, write to them at 3280 South Miami Avenue, Miami, Florida 33129 for guidance in helping to locate unusual fruiting trees and plants.

The Dade County Cooperative Extension Service in Homestead has dozens of free booklets and folders on the care of many tropical fruit trees and will answer letters and questions by phone on all phases of tropical fruit tree culture.

PLANTING GUIDE FOR VEGETABLE GARDENS

Florida Cooperative Extension Service, Institute of Food and Agricultural Sciences, University of Florida, Gainesville

Crop	Varieties	Planting Dates in Florida (incl.)			Plant Hardi-ness†
		North	Central	South	
Beans, Snap	Extender, Contender, Harvester, Wade, Cherokee (wax)	Mar.-Apr. Aug.-Sept.	Feb.-Mar. Sept.	Sept.-Apr.	T
Beans, Pole	Dade, McCaslan, Kentucky Wonder 191, Blue Lake	Mar.-June	Feb.-Apr.	Jan.-Feb.	T
Beans, Lima	Fordhook 242, Concentrated, Henderson, Jackson Wonder, Dixie Butterpea, Florida Butter (Pole)	Mar.-June	Feb.-Apr.	Sept.-Apr.	T
Beets	Early Wonder, Detroit Dark Red	Sept.-Mar.	Oct.-Mar.	Oct.-Feb.	H
Broccoli	Early Green Sprouting, Waltham 29, Atlantic	Aug.-Feb.	Aug.-Jan.	Sept.-Jan.	H
Cabbage	Copenhagen Market, Marion Market, Badger Market, Glory of Enkhuizen, Red Acre, Chieftan Savoy	Sept.-Feb.	Sept.-Jan.	Sept.-Jan.	H
Cantaloupe	Smith's Perfect, Seminole, Edisto 47, Gulfstream	Mar.-Apr.	Feb.-Apr.	Feb.-Mar.	T

PLANTING GUIDE FOR VEGETABLE GARDENS (Contd.)

Crop	Varieties	Planting Dates in Florida (incl.) North	Central	South	Plant Hardi-ness†
Carrots	Imperator, Gold Spike, Chantenay, Nantes	Sept.-Mar.	Oct.-Mar.	Oct.-Feb.	H
Cauli-flower	Snowball Strains	Jan.-Feb. Aug.-Oct.	Oct.-Jan.	Oct.-Jan.	H
Celery	Utah 52-70, Florida	Jan.-Mar.	Aug.-Feb.	Oct.-Jan.	H
Chinese	Michihli, Wong Bok	Oct.-Jan.	Oct.-Jan.	Nov.-Jan.	H
Chinese Cabbage	Michihli, Wong Bok	Oct.-Jan.	Oct.-Jan.	Nov.-Jan.	H
Collards	Georgia, Vates	Feb.-Mar.	Jan.-Apr. Sept.-Nov.	Sept.-Jan. Aug.-Nov.	H
Corn, Sweet	Silver Queen (white), Gold Cup, Golden Security, Seneca Chief, many others	Mar.-Apr.	Feb.-Mar.	Jan.-Feb.	T
Cucum-bers	Poinsett, Ashley (slicers), Wisconsin SMR 18, Pixie (picklers)	Feb.-Apr.	Feb.-Mar. Sept.	Jan.-Feb.	T
Eggplant	Florida Market	Feb.-Mar. July	Jan.-Feb. Aug.-Sept.	Dec.-Feb.	T
Endive-Escarole	Deep Heart Fringed, Full Heart Batavian	Feb.-Mar. Sept.	Jan.-Feb. Sept.	Sept.-Jan.	H
Kohlrabi	Early White Vienna	Mar.-Apr. Oct.-Nov.	Feb.-Mar. Oct.-Nov.	Nov.-Feb.	H
Lettuce (Crisp) (Butter-head) (Leaf) (Romaine)	Premier, Great Lakes types, Bibb, Matchless, Sweetheart, Prize Head, Ruby, Salad Bowl, Paris Island Cos, Dark Green Cos	Feb.-Mar. Sept.	Jan.-Feb. Sept.	Sept.-Jan.	H
Mustard	Southern Giant Curled, Florida Broad Leaf	Jan.-Mar. Sept.-May	Jan.-Mar. Sept.-Nov.	Sept.-Mar.	H
Okra	Clemson Spineless, Perkins Long Green	Mar.-May Aug.	Mar.-May Aug.	Feb.-Mar. Aug.-Sept.	T

PLANTING GUIDE FOR VEGETABLE GARDENS (Cont'd.)

Crop	Varieties	Planting Dates in Florida (incl.) North	Central	South	Plant Hardiness†
Onions (Bulbing)	Excel, Texas Grano, Granex, White Granex,	Jan.-Mar. Aug.-Nov.	Jan.-Mar. Aug.-Nov.	Jan.-Mar. Sept.-Nov.	H
(Green)	Tropicana Red White	Aug.-Mar.	Aug.-Mar.	Sept.-Mar.	H
	Portugal or White types, Shallots (Multipliers)	Aug.-Jan.	Aug.-Jan.	Sept.-Dec.	H
Parsley	Moss Curled, Perfection	Feb.-Mar.	Dec.-Jan.	Sept.-Jan.	H
Peas	Little Marvel, Dark Skinned Perfection, Laxton's Progress	Jan.-Feb.	Sept.-Mar.	Sept.-Feb.	H
Peas Southern	Blackeye, Brown Crowder, Bush Conch, Producer, Floricream, Snapea, Zipper Cream	Mar.-May	Mar.-May	Feb.-Apr.	T
Pepper (Sweet)	Calif. Wonder, Yolo Wonder, World Beater	Feb.-Apr.	Jan.-Mar.	Jan.-Feb. Aug.-Oct.	T
(Hot)	Hungarian Wax, Anaheim Chili				
Potatoes	Sebago, Red Pontiac, Kennebec, Red LaSoda	Jan.-Feb.	Jan.	Sept.-Jan.	SH
Potatoes, Sweet	U.S. No. 1, Porto Rico, Georgia Red, Goldrush, Nugget, Centennial	Mar.-June	Feb.-June	Feb.-June	T
Radish	Cherry Belle, Comet, Early Scarlet Globe, White Icicle, Sparkler (white tipped)	Oct.-Mar.	Oct.-Mar.	Oct.-Mar.	H
Spinach	Virginia Savoy, Dixie Market, Hybrid 7	Oct.-Nov.	Oct.-Nov.	Oct.-Jan.	H
Spinach, Summer	New Zealand	Mar.-Apr.	Mar.-Apr.	Jan.-Apr.	T
Squash (Summer)	Early Prolific Straightneck, Early Summer Crookneck, Cocozelle, Zucchini, Patty Pan	Mar.-Apr. Aug.	Feb.-Mar. Aug.	Jan.-Mar. Sept.-Oct.	T

PLANTING GUIDE FOR VEGETABLE GARDENS (Cont'd.)

Crop	Varieties	Planting Dates in Florida (incl.)			Plant Hardi-ness†
		North	Central	South	
Squash (cont'd.)					
(Winter)	Alagold, Table Queen, Butternut	Mar.	Feb.-Mar.	Jan.-Feb.	T
Straw-berry	Florida 90, Tioga, Sequoia	Sept.-Oct.	Sept.-Oct.	Oct.-Nov.	H
Tomatoes (Large Fruited)	Manalucie,‡ Home-stead-24, Indian River, Floradel,‡ Tropired, Big Boy‡ Walter,	Feb.-Apr. Aug.	Feb.-Mar. Sept.	Aug.-Mar.	T
(Small Fruited)	Large Cherry, Roma (Paste)	Feb.-Apr. Aug.	Feb.-Mar. Sept.	Aug.-Mar.	T
Turnips	Japanese Foliage (Shogoin) Purple Top White Globe	Jan.-Apr. Aug.-Oct.	Jan.-Mar. Sept.-Nov.	Oct.-Feb.	H
Water-melon (Large)	Charleston Gray, Congo, Jubilee, Crimson Sweet	Mar.-Apr.	Jan.-Apr.	Feb.-Mar.	T
(Seedless)	Tri-X 317				
(Small)	New Hampshire Midget, Sugar Baby				

Other Vegetables for the Garden.—Jerusalem artichoke, Brussels sprouts, cassava, chayote, chives, dandelion, dasheen, dill, fennel, garbanzo bean, garlic, herbs, kale, leek, luffa gourd, honeydew melons, and rutabaga.

Note—globe artichokes, asparagus, and rhubarb not well adapted to Florida.

†H - Hardy, can stand frost and usually some freezing (32°F) without injury.

SH - Slightly hardy, will not be injured by light frosts.

T - Tender, will be injured by light frost.

‡ - Tomato varieties best adapted to staking.

SOWING AND HARVESTING GUIDE

Florida Cooperative Extension Service, Institute of Food and Agricultural Sciences, University of Florida, Gainesville

Crop	Seed/Plants 100' of Row	Spacing in Inches		Seed Depth Inches	Pounds Yield 100'	Days To Harvest
		Rows	Plants			
Beans, Snap	1 lb.	18-30	2-3	1½-2	45	50-60
Beans, Pole	1 lb.	40-48	15-18	1½-2	80	60-65
Beans, Lima	1 lb.	26-48	12-15	1½-2	50	65-75
Beets	1 oz.	14-24	3-5	½-1	75	60-70
Broccoli	60 plts. (¼ oz.)	30-36	16-22	½-1	50	60-70
Cabbage	65 plts. (¼ oz.)	24-36	14-24	½	125	70-90
Cantaloupes	1 oz.	70-80	48-60	¾	150	75-90
Carrots	½ oz.	16-24	1-3	½	100	70-75
Cauliflower	55 plts. (¼ oz.)	24-30	20-24	½	80	55-60
Celery	150 plts. (¼ oz.)	24-36	6-10	¼-½	150	115-125
Chinese Cabbage	125 plts. (¼ oz.)	24-36	8-12	¼-½	100	75-85
Collards	75 plts. (¼ oz.)	24-30	14-18	½	150	50-55
Corn, Sweet	¼ lb.	34-42	12-18	½	15	80-85
Cucumbers	1 oz.	48-60	15-24	½-¾	100	50-55
Eggplant	30 plts. (¼ oz.)	36-42	36-48	½	200	80-85
Endive-Escarole	1 oz.	18-24	8-12	¾	75	90-95
Kohlrabi	½ oz.	24-30	3-5	½	100	50-55
Lettuce	½ oz.	12-18	12-18	¾	75	50-80
Mustard	1 oz.	14-24	4-8	½	100	40-45
Okra	2 oz.	24-40	18-24	1-2	70	50-55
Onions (Bulbing)	400 plts. or sets 1 oz. seed	12-24	3-4	¾	100	100-300
(Green)	800 plts. or sets	12-24	1½-2	¾	100	50-75
	1½ lb. seed	18-24	6-8	¾	100	75-105

SOWING AND HARVESTING GUIDE (Cont'd.)

Crop	Seed/Plants 100' of Row	Spacing in Inches		Seed Depth Inches	Pounds Yield 100'	Days To Harvest
		Rows	Plants			
Parsley	1 oz.	12-20	8-12	¾	40	90-95
Peas	1½ lbs.	24-36	2-3	1-2	40	50-55
Peas, Southern	1½ lbs.	30-36	2-3	1-2	80	70-80
Pepper	60 plts. (¼ oz.)	20-36	18-24	½	50	70-80
Potatoes	15 lbs.	36-42	12-15	4-8	150	80-95
Potatoes, Sweet	80 plts.	48-54	18-24	—	75	120-140
Radish	1 oz.	12-18	1-2	¾	40	20-25
Spinach	2 oz.	14-18	3-5	¾	40	40-45
Spinach, Summer	2 oz.	30-36	18-24	¾	40	55-65
Squash						
(Summer)	2 oz.	42-48	42-48	½	150	45-60
(Winter)	2 oz.	90-120	48-72	2	300	95-105
Strawberry	100 plts.	36-40	10-14	—	50	90-110
Tomatoes (Large Fruited)	35 plts. (¼ oz.)	40-60	36-40	½	125	75-85
(Small Fruited)	70 plts. (½ oz.)	36-48	18-24	½	200	75-85
Turnips	1 oz.	12-20	4-6	½-¾	150	40-50
Watermelon	2 oz.	90-120	60-84	2	400	80-100

APPROXIMATE AMOUNT OF SEED AND SPACE NEEDED TO FEED AVERAGE FAMILY

Kind of vegetable	Number running feet of row for a family of 3 or 4 people (Includes for canning)	Seeds or plants to provide for a family of 3 or 4 people
Beans (Bush)	100*	1/2 lb.†
Beans (Pole)	50	1/4 lb.
Beans (Bush Lima)	100*	1/2 to 1 lb.†
Beets	100*	1 oz.†
Broccoli	40	1 pkt. or 25 plts.
Brussels Sprouts	25	15 plants
Cabbage	60	50 plants
Chinese Cabbage	40*	1 pkt. or 50 plts.†
Cantaloupe	50	1 pkt.†
Carrots	100*	2 pkts.†
Cauliflower	50	30 plants
Celery	50	150 plants
Collards	50*	1 pkt. or 35 plts.†
Corn (Sweet)	200*	1/4 lb.†
Corn (Roasting Ear)	200*	1/4 lb.†
Cowpeas (Table)	150	1/2 to 3/4 lb.
Cucumbers	50	1 pkt.
Eggplants	50	17 plants
Endive (Escarole)	40	pkt. or 50 plts.
Kale	25	1 pkt.
Kohlrabi	25*	1 pkt.†
Leek	50	1 pkt. or 120 plts.
Lettuce (Head)	75*	1 pkt. or 90 plts.†
Lettuce (Leaf)	50*	1 pkt. or 60 plts.†
Mustard	40*	1 pkt.†
Okra	75	1 oz.
Onion Seed	80	1 pkt. or 400 plts.
Onion Sets	40	1 pt.
Parsley	30	1 pkt.
Parsnip	50	1 pkt.
Peas (English)	100*	1/2 lb.†
Peppers	40	24 plants
Potatoes (Irish)	200	12 lbs.
Pumpkins	50	1/2 oz.
Radishes	25*	1 pkt.†
Rhubarb	25	1 pkt. or 15 plts.
Rutabagas	50	1 pkt.
Spinach - Savoy	50*	1 pkt.†
Squash (Bush)	50	1 pkt.
Squash (Running)	50	1 pkt.
Swiss Chard	25	1 pkt.
Tomatoes (Ground)	150	50 plants
Tomatoes (Staked)	75	50 plants
Turnips	50	1 pkt.†
Watermelons	75	1 pkt.

*Make two or more plantings at different times during the season. Number of feet or row indicated is for each planting.

†No. of ft. and am't. of seed or no. of plts. indicated are for each planting.

AVERAGE PLANT FOOD CONTENT OF NATURAL AND ORGANIC FERTILIZER MATERIALS (PERCENTAGE ON A DRY-WEIGHT BASIS)

Organic Materials	%Nitrogen	%Phosphate	%Potash	Availability	Acidity
Fish Scrap	5.0	3.0	0	slowly	acid
Fish Meal	10.0	4.0	0	slowly	acid
Guano, Peru	13.0	8.0	2.0	moderately	acid
Guano, Bat	10.0	4.0	2.0	moderately	acid
Sewage Sludge	2.0–6.0	1.0–2.5	0.0–0.4	slowly	acid
Dried Blood	12.0	1.5	0.8	mod. slow	acid
Soybean Meal	7.0	1.2	1.5	slowly	v. sl. acid
Tankage, Animal	9.0	10.0	15.5	slowly	acid
Tankage, Garbage	2.5	1.5	1.5	very slowly	alkaline
Tobacco Stems	1.5	0.5	5.0	slowly	alkaline
Seaweed	1.0	–	4.0–10.0	slowly	–
Bone Meal, Raw	3.5	22.0	–	slowly	alkaline
Urea	45.0	–	–	quickly	acid
Caster Pomace	6.0	1.2	0.5	slowly	acid
Wood Ashes	–	2.0	4.0–10	quickly	alkaline
Coco Shell Meal	2.5	1.0	2.5	slowly	neutral
Cotton Seed Meal	6.0	2.5	1.5	slowly	acid
Ground Rock Phosphate	–	33.0	–	very slowly	alkaline
Green Sand	–	1.0	6.0	very slowly	–
Basic Slag	–	8.0	–	quickly	alkaline
Horn and Hoof Meal	12.0	2.0	–	–	–
Milorganite	6.0	2.5	–	–	–
Peat and Muck	1.5–3.0	0.25–0.5	0.5–1.0	very slowly	acid

NOTE: Urea and calcium cyanamide are organic compounds, but since they are synthetic, it is doubtful that most organic gardeners would consider them acceptable.

COMPOSITION OF VARIOUS MATERIALS USED IN COMPOST PILES

Compost Material	%Nitrogen	%Phosphate	% Potash
Banana Skins (Ash)	—	3.25	41.76
Cantaloupe Rinds (Ash)	—	9.77	12.21
Castor Bean Pomace	5.00	2.00	1.00
Cattail Reeds	2.00	.81	3.43
Coffee Grounds	2.08	.32	.28
Corncob Ash	—	—	50.00
Corn Stalks and Leaves	.30	.13	.33
Crabgrass, Green	.66	.19	.71
Eggs, Rotten	2.25	.40	.15
Feathers	15.30	—	—
Fish Scraps	2.0–7.50	1.5–6.0	—
Grapefruit Skins (Ash)	—	3.58	30.60
Oak Leaves	.80	.35	.15
Orange Culls	.20	.13	.21
Pine Needles	.46	.12	.03
Ragweed	.76	.26	—
Tea Grounds	4.15	.62	.40
Wood Ashes	—	1.00	4.0–10.00

HARVEST PERIODS FOR FRUIT AND VEGETABLE CROPS IN DADE COUNTY

VEGETABLE	OCT.	NOV.	DEC.	JAN.	FEB.	MAR.	APR.	MAY.	JUNE.	JULY.	AUG.	SEPT.
BEAN, BUSH GREEN	■	■	■	■	■	■	■					
BEANS, POLE	■	■	■	■	■	■	■					
BEETS			■	■	■	■						
CABBAGE			■	■	■	■	■	■				
CORN, SWEET					■	■	■	■				
CUCUMBERS	■	■	■	■	■	■	■	■	■			■
EGGPLANT	■	■	■	■	■	■	■	■				
OKRA	■	■	■		■	■	■	■	■	■	■	■
PEAS, SOUTHERN	■	■	■				■	■	■	■	■	■
POTATOES, IRISH					■	■	■					
SQUASH	■	■	■	■	■	■	■	■	■	■		■
STRAWBERRIES			■	■	■	■	■					
TOMATOES		■	■	■	■	■	■					

SUBTROPICAL FRUITS	OCT.	NOV.	DEC.	JAN.	FEB.	MAR.	APR.	MAY.	JUNE.	JULY.	AUG.	SEPT.
AVOCADOS	■	■	■	■	■	■		■	■	■	■	■
LIMES	■	■	■	■	■	■	■	■	■	■	■	■
MANGOS								■	■	■	■	■

Distributed by: Dade County Extension Home Economics Offices

For free booklets and answers to questions on vegetable growing and tropical fruit raising:

Dade County Cooperative Extension Service, 18710 S.W. 288 Street, Homestead, Florida 33030 (phone 248-3311). Agent-Vegetables: Dr. William Stall; Agent-Fruit Crops: Mr. Seymour Goldweber. Miami office: 2690 N.W. 7 Avenue, Miami 33127.

"Market Bulletin": Florida Department of Agriculture and Consumer Services, Mayo Building, 407 Calhoun Street, Tallahassee, Florida 32004. Free subscriptions, free advertising.

Hydroponics information: Dr. Chatelier's Plant Food Co., P.O. Box 20375, St. Petersburg, Florida 33742

Seed companies:

Burgess Seed and Plant Co., P.O. Box 218, Galesburg, Michigan 49053

W. Atlee Burpee Co., Philadelphia, Pennsylvania 19132

Kilgore Seed Co., 1400 W. 1 Street, Sanford, Florida 32771

George W. Park Co., Inc., Greenwood, South Carolina 29646

R. H. Shumway, Seedsman. 628 Cedar Street, Rockford, Illinois 61101

Other:

Rare Fruit Council International, Inc., 3280 S. Miami Avenue, Miami, Florida 33129

Fairchild Tropical Garden, 10901 Old Cutler Road, Coral Gables, Florida 33156, has many interesting courses, including one on vegetable gardening. Phone 667-1651 for information.

For Dade County's "U-Pick-Em" bargain fields to supplement homegrown produce:

Follow Old Cutler Road on the east, North Kendall Drive to its limit on the north, and Krome Avenue on the west. West of Krome, follow 232 Street, 168 Street, 256 Street, and other country lanes.

Books and publications useful to tropical growers:

*Agricultural Extension Service, Univ. of Florida (Gainesville, Fla.). "Canning Florida Fruits and Vegetables." Bulletin 155.

Dade County Cooperative Extension Service. "Fruit Plants for South Florida."

*Dade County Cooperative Extension Service. "Organic Vegetable Gardening." Circular 375.

*Dade County Cooperative Extension Service. "Recipes for Dade Grown Boniato."

Dorn, Mabel. *Under the Coconuts.* Privately published and out of print but available in some public libraries.

Doty, Walter L., ed. *All About Vegetables–South Edition.* Chevron Chemical Co., Ortho Division.

*Durre, Nolan L. "Vegetable Gardening Guide."

*Durre, Nolan L. "Vegetable Garden Notes." Lists of vegetables and their pests, diseases, and controls by the former Dade County Cooperative Extension Service Agent-Vegetables.

Maxwell, Lewis. *Florida Insects.* Tampa, Fla.: Lewis Maxwell (6340 Travis Blvd., Tampa 33610).

Maxwell, Lewis. *Florida Vegetables.* Tampa, Fla.: Lewis Maxwell (6340 Travis Blvd., Tampa 33610).

Morton, Julia F. *Wild Plants for Survival in South Florida.* Tampa, Fla.: Trend House.

Perry, Ann M. *Dooryard Supermarket in the Tropics and Subtropics.* Published by the author; out of print but available at some public libraries.

Purseglove, J. W. *Tropical Crops.* 4 vols. London: Longmans, Green and Company, Ltd.

*Stall, William. "Vegetable Gardening in Dade County."

Sturtevant, E. Lewis. *Sturtevant's Notes on Edible Wild Plants.* Ed. by U. P. Hedrick. New York: Dover Publications, Inc.

Sunset Books, eds. *Gardening in Containers.* Menlo Park, Calif.: Lane Magazine and Book Co.

Sunset Books, eds. *Vegetable Gardening.* Menlo Park, Calif.: Lane Magazine and Book Co.

Time-Life Encyclopedia of Gardening, *Vegetables and Fruits.* New York: Time-Life Books, Time Inc. Contains Alexis Bervey's blueprints for a Miami vegetable garden.

*U.S. Department of Agriculture. "Home Freezing of Fruits and Vegetables." Home and Garden Bulletin No. 10. Washington, D.C.: Government Printing Office.

*U.S. Department of Agriculture. "Insects and Diseases of Vegetables in the Home Garden." Home and Garden Bulletin No. 46. Washington, D.C.: Government Printing Office.

*Available through Dade County Cooperative Extension Service, Homestead and Miami.

Index